MYCOLOGIST'S HANDBOOK

This and other publications of the
Commonwealth Agricultural Bureaux
can be obtained through any major bookseller or
direct from
Commonwealth Agricultural Bureaux,
Central Sales, Farnham Royal, Slough, SL2 3BN, England

MYCOLOGIST'S HANDBOOK

An Introduction to the Principles of Taxonomy and Nomenclature in the Fungi and Lichens

by

D. L. HAWKSWORTH
B.Sc., Ph.D., F.L.S.

Mycologist,
Commonwealth Mycological Institute,
Kew

COMMONWEALTH MYCOLOGICAL INSTITUTE
KEW, SURREY, ENGLAND

1974

First published in June 1974 by the Commonwealth Mycological Institute
under the authority of the
Executive Council, Commonwealth Agricultural Bureaux,
Farnham Royal, Bucks., England

Reprinted 1977

ISBN 0 85198 300 6 (cased)

ISBN 0 85198 306 5 (paperback)

Printed in Great Britain by
Western Printing Services Ltd, Bristol

TO G. C. AINSWORTH IN RECOGNITION
OF HIS SERVICES TO MYCOLOGY

PREFACE

THIS book owes its inspiration to the late Dr G. R. Bisby's *An Introduction to the Taxonomy and Nomenclature of Fungi*, first published by this Institute in 1945 (with a second edition issued in 1953). By the end of 1969 stocks of the second edition of Bisby's book were almost exhausted and it was clear that a new edition or replacement volume would be required. To both bring the work up to date and provide a useful reference volume for all those concerned with the naming of fungi, the entire text has been completely re-written, many new chapters included, and a few omitted.

Bisby, in common with most of his contemporaries, did not include data relevant to over half of all known Ascomycotina (i.e. the lichen-forming species) and some attempt to rectify this situation has been made.

The portions of the *International Code of Botanical Nomenclature* pertinent to mycologists and illustrated by examples from mycology were originally included by Bisby because the 1935 edition of the Code was not readily available. Although this is no longer the case, the Code has necessarily increased in complexity to such an extent that there is perhaps now a greater need for a Code with mycological examples than there was in 1945.

In a work of this size it is, of course, impossible to treat all the practical and theoretical aspects of the nomenclature and taxonomy of fungi (including the lichens) in the detail they merit and consequently many references to books and papers from which further information can be obtained have been included.

I would like to thank some of my colleagues for their kind assistance in reading and commenting on drafts of various sections: in particular Dr C. Booth, Miss S. Daniels, Mr F. C. Deighton, Mr D. W. Fry, Mr A. Johnston, Dr A. H. S. Onions and Dr B. C. Sutton, all at the CMI; and Mr P. W. James, Mr J. R. Laundon, Dr F. Rose and the mycologists at the Plant Research Institute, Ottawa. I am also grateful to Dr F. A. Stafleu and the International Association for Plant Taxonomy for permission to reproduce portions of the *International Code of Botanical Nomenclature*, and to the following for permission to reproduce line-figures: The Botanical Society of the British Isles and Dr W. T. Stearn (Fig. 17), the British Lichen Society (Figs. 14, 16, 18) and Royal VanGorcum Ltd (Fig. 20).

In conclusion I would like to emphasize that any errors and omissions are my own responsibility.

D.L.H.

Commonwealth Mycological Institute
Kew
15 May 1973

CONTENTS

I. INTRODUCTION

THIS book is intended for all who are concerned with the naming of fungi. It aims to outline the principles and techniques involved in the naming and describing of fungi and lichens, in undertaking taxonomic and floristic studies, and in the formation and maintenance of herbaria and culture collections. The rules controlling the nomenclature of fungi and lichens (The International Code of Botanical Nomenclature), illustrated by examples taken from mycology, together with a glossary of nomenclatural terms, are included. Notes on the application of some recent techniques of value to systematists are also provided as are guides to the tracing of literature sources, the location of some important mycological collections, preparing manuscripts for publication and suggested title abbreviations for mycological books.

Accounts of the different groups of fungi and works to be consulted for the determination of genera and species are outside the scope of this volume. These data are readily obtainable through the latest (sixth) edition of *Ainsworth & Bisby's Dictionary of the Fungi* (Ainsworth, 1971), which includes the lichens, and Ainsworth *et al.* (1973). As general introductions the texts of Bessey (1950), Alexopoulos (1962), Burnett (1968), Webster (1970) and Talbot (1971) are recommended for the fungi; whilst those of Ahmadjian (1967a), Hale (1967) and Henssen and Jahns (1974) are for the lichens. A résumé of recent advances in the study of lichens is given by Hawksworth (1973a). More detailed accounts of most aspects of fungi and lichens are provided in the comprehensive works of Ainsworth and Sussman (1965, 1966, 1968) and Ahmadjian and Hale (1973*), respectively.

A great deal of revisionary work still needs to be carried out in many groups of fungi and lichens and, although they have been studied for over two centuries, their taxonomy largely remains at a stage of 'cataloguing' with new genera and species continually being described. Revisionary studies are consequently becoming increasingly necessary to draw together the large numbers of species already known.

Systematic mycology is not a subject the study of which can be undertaken lightly. There is a vast literature, and microscopic characters have to be considered for all species, some of which are invisible to the unaided eye. Furthermore the student must have a general systematic biological background for it is essential to determine accurately the hosts and substrates on which fungi occur. It is because of these inherent difficulties and intricacies, however, that systematic mycology can be particularly rewarding for the persevering and careful student.

There are too many fungi and lichens for any individual to study all groups

---

* This work appeared too late in 1973 to enable data from it to be fully incorporated into the text.

in detail and so the student has to decide in what area he wishes to work. The amateur has complete freedom to study whatever groups may interest him but someone who hopes to make his career in this field has to be more selective. The choice of such a person might be governed by the work of colleagues in his Department or Institute, or by the pathological or industrial importance of particular genera and species. To avoid the duplication of work being carried out elsewhere the potential mycologist should discuss the area in which he proposes to work with as many professional mycologists as possible. Specialists in particular groups are usually aware of studies in progress in other laboratories and herbaria long before they appear in print and are also able to direct others to groups where systematic work is required and where the problems are not likely to prove insurmountable. If such advice cannot be obtained some idea of the studies in progress in particular groups may be obtained from Bulletins, Newsletters and publications appearing on the group (pp. 107–108).

The types of study within the field of systematic mycology which it is possible to carry out may also be limited by the facilities available (e.g. libraries, herbaria, culture collections, good microscopes, chromatography apparatus). Amateurs are able to make important contributions by sending interesting collections to specialists and compiling lists of species present in particular areas, or on specific hosts or substrates. Specialists determining material sent to them are often more inclined to give advice and name specimens if they know that the information they provide is being included in a definite project.

Whatever aspect of systematic mycology is pursued, patience, persistence, care and adequate facilities are prerequisites. The student must not be too eager to revise taxa he really knows little about, nor to describe new genera and species without making certain that they have not been described before. Once a name has been validly published it is irrevocably introduced and requires cataloguing and discussion by other mycologists even if it is found to be the same as a previously described taxon. The publication of a new taxon is consequently a step which should not be undertaken lightly. It is the intention of this book both to help the student to make 'correct' taxonomic decisions and to indicate how best to study and preserve his material and publish his results.

II. COLLECTION AND PRESERVATION

COLLECTION

BEFORE a beginner can take a real interest in systematic mycology and appreciate its problems he needs to acquire a general knowledge of the various groups of fungi to provide a background for further discussion and study. The most satisfactory way to do this is to learn the common species in his own neighbourhood and, to do this, he must collect and study them. He should attempt to develop a keen eye so that smaller and critical species are not overlooked, learn the types of habitats in which particular groups predominate, and recognize the range of variation within individual species in his area. Systematic collecting provides the basis of all taxonomic research. Someone working on a particular group in the laboratory or herbarium also needs to study and collect material of his group in the field. Only through field studies can he become familiar with the inherent variability of species and the effects of environmental factors on them.

The ideal way in which to learn how to collect material is to join field courses, field excursions and forays of national and local natural history societies led by specialists. On such meetings more can often be learnt in an afternoon than an individual working on his own could hope to learn in several weeks. Most collectors at first bring back specimens which are either too small or in too poor a condition for accurate identification. Careful searching of a very limited number of sites is often more fruitful than briefer visits to a large number. Both the quality of the collections obtained and the number of species found tend to be in inverse proportion to the distance travelled whilst collecting.

All collectors should obtain permission to collect in an area from its owner where the property is private. Failure to do this may jeopardize visits by other naturalists (not only mycologists) in the future. As future generations will also wish to study fungi and lichens the collector should be conscious of the need for conservation. Most ephemeral fungi are unlikely to be affected by overcollecting but some groups, particularly the lichens, may all too easily be endangered by overzealous collectors. Lichenologists themselves may have contributed to the decline to extinction of a few rare species in the British Isles. Fortunately, with practice, most lichens can be named in the field with a hand lens ($\times$10) and where voucher material is needed or a specimen requires microscopic examination only the minimum of material should be taken. Later workers have often been sceptical of unsupported reports of rare lichens they have been unable to refind in the same localities. Collections should reflect what grows in a locality and not what used to grow there before the collector visited it. In collecting species growing on bark the minimum amount of damage to the tree should be made and where possible the living tissues of the tree not penetrated.

Accurate determinations of hosts and substrates are always necessary. If there is some uncertainty as to the identity of a host plant adequate material for its identification (i.e. flowers, fruits, leaves) should also be collected. Species are sometimes reported from incorrectly named hosts and collectors should ensure that the possibility of this occurring as a result of their slips is remote. Specimens from particular hosts or substrates in the same locality may conveniently be placed in separate carefully labelled bags. White card or paper slips with the locality and host indicated on them in pencil (ink and biro are not satisfactory as they tend to run and become illegible if the material is damp) may then be inserted into each bag. If paper bags or envelopes are employed these data may be written directly on them but polythene bags or tins are more suitable for collecting damp material. Larger fleshy fungi are likely to be damaged if placed in bags and should be wrapped individually in newspaper and carefully placed in flat-bottomed wicker market-baskets.

If material cannot be examined the same day it is collected it should not be left in a damp condition in air tight containers such as polythene bags but thoroughly dried to prevent the growth of unwanted moulds (saprophytic fungi). Notes on any characters likely to be affected by drying (see below) must be made and placed with the specimens as soon as possible.

The beginner is apt to attempt to collect at random anything which catches his eye but this approach tends to lead to very superficial lists from an area. The mycologist should endeavour to pay particular attention to clearly delimited habitats, hosts and groups, on different visits to a site at as many times of the year as possible, if he wants to acquire a reasonably comprehensive knowledge of the fungi in an area. Most mycologists find it extremely difficult to search for all groups in all habitats at the same time.

Some notes on collecting different groups of fungi and techniques for studying some special habitats are given below. It is not possible to discuss all of these in detail here and the references cited should be consulted for further information. A comprehensive account of procedures for collecting and examining fungi of different groups is currently being prepared by the Mycological Society of America. General accounts of collecting and preservation procedures are provided in the British Museum (Natural History)'s *Instructions for Collectors* no. 10 *Plants* (ed. 6, 1957) and by Savile (1962), Fosberg and Sachet (1965) and Smith (1971).

Fleshy fungi

Fleshy fungi (together with larger woody species often now referred to as 'macromycetes'), particularly the larger members of the Agaricales, are the easiest fungi to observe and at the same time one of the most difficult groups to preserve satisfactorily. Whenever possible several healthy specimens of each species at different stages of development should be collected by digging (*not* pulling) them up so as not to damage their bases. On returning from a collecting trip the material should be examined on the same day and, if it is not possible to name the specimens, the information on characters likely to be affected by drying must be obtained so that they can be determined later. Habit sketches, measurements, colour of different parts, colour on bruising and cutting the flesh, any exudates on cutting, odour, consistency, spore-prints, and any chemical tests (where appropriate; see e.g. Zoberi, 1972) should all be made at this stage. Accurate descriptions of colour are particularly important and some agaricologists paint their sketches using water-colour paints; 35 mm coloured photographic slides are particularly helpful in this connection. Because of the time needed to obtain this information more material than can be properly documented in the available time should not be collected.

When specimens have been documented properly they can be dried rapidly (e.g. over a radiator) and microscopic characters are usually unaffected by this procedure although the colour, shape and size may change dramatically. A useful technique for preparing herbarium specimens of fleshy fungi by slicing them vertically and drying between blotting paper is described by Bohus (1963).

The most satisfactory method of preserving fleshy fungi, however, is by freeze-drying (lyophilization) as specimens preserved in this way show little change in either colour or shape. Freeze-dried specimens are often fragile and require careful handling although this may be overcome to some extent by dipping them in a polyurethane (2):white spirit (1) solution and drying at 50–60°C (see Onions, 1971; Kendrick, 1969). A method of preparing models of fleshy fungi in epoxy resin for display purposes is described by Parmelee (1971).

Rusts and smuts

Rusts and smuts are amongst the easiest fungi to preserve. Rusts should be searched for on both sides of leaves and smuts in the flowers or on herbaceous stems and fruits which are splitting open. Careful determination of hosts is essential in these groups as identification of the fungus may prove difficult if the host is not known or incorrectly named. In collecting rusts search for teliospores as well as the orange-yellow urediniospores ('uredospores'); teliospores form in discrete patches similar to those forming urediniospores but are brown to black. Rusts and smuts are readily preserved by drying the infected leaves between sheets of newspaper under light pressure.

Ascomycotina

Larger fleshy Ascomycotina (e.g. species of *Morchella* Dill. ex Fr. and *Peziza* Dill. ex Fr.) require treatment as described above for fleshy fungi generally (pp. 14–15) but most are readily preserved by drying them slowly. Many species occur on wood and to collect these a strong sheath knife, secateurs and small saw are required. In some pyrenomycetes the perithecial stromata can persist long after the spores have been discharged. To avoid collecting effete material, which is usually unnamable, slice across the edge of a stroma or the top of a few perithecia; if the locules are producing spores they will be seen with the aid of a hand lens ($\times 10$) to be filled with whitish mucilage or have a shiny smooth interior but if the fungus is dead the inner walls of the locules appear dull and mucilage is absent.

Many small ascomycetes grow just below the surfaces of stems, leaves and bark with only their minute ostioles protruding. These are easily overlooked without careful searching but species occuring beneath bark can often be discovered by peeling back the surface layers of the bark with a sharp finger-nail or knife. Dung, straw, dead leaves, twigs and stems are also important habitats to search for ascomycetes. Damp rotting stems and decaying wood often have many small fleshy discomycetes.

Lichens

Corticolous, foliicolous and terricolous lichens are collected by similar methods to those employed in other fungi. When foliose lichens are being collected attempts should not be made to peel them from their substrates as this can damage rhizinae which may be needed to identify them; basal parts of some fruticose lichens also have some diagnostic value and so these should not be torn from their substrates but removed carefully paying particular attention to securing their bases. All specimens collected should be representative of the population from which they are taken. Many species of corticolous lichens often occur on single trees (over 30 species frequently occur on some trees in areas of Britain relatively free from air pollution by sulphur dioxide) and a detailed study of a single tree (particularly bark fissures, tree bases and twigs) with a hand lens ($\times 10$) may be needed to detect many of the smaller species. Lichenologists often spend 20–30 minutes studying an individual tree.

Saxicolous lichens (i.e. lichens growing on rocks) are collected with the aid of a geological hammer (1–2 lb) and small cold-steel sharp masonry chisels ($\frac{1}{4}$ and $\frac{1}{2}$ in). Some rocks are difficult to collect from as they tend to flake and disintegrate on chiselling and in some cases it is necessary to scrape material off with a sharp knife. Tombstones and privately owned walls should not be chiselled and in examining these it is most satisfactory to carry out chemical reagent tests and make slide preparations and notes in the field. Rock samples need to be individually wrapped in tissue paper to prevent their being damaged by rubbing against other samples in the collectors' bag. Some species are restricted to particular rock types and so it is important to examine as many rock types as may ·be present in an area. Before making field excursions it is consequently valuable first to study a geological map of the area to be visited.

In collecting crustose species on soil or in rock crevices the soil immediately below the material should be removed with the specimen, disturbed as little as possible, and carefully wrapped in tissue paper; it can be very tiresome trying to find ascocarps among soil powder in the corner of a paper packet on returning to the laboratory (see also p. 25).

Lichens of all groups are easily preserved by drying (e.g. over a radiator) and show little change (apart from colour in some instances).

For further information on collecting lichens see Duncan (1970).

Deuteromycotina

The collection of Coelomycetes and Hyphomycetes requires careful searching of dead and living stems, bark and leaves. The procedures involved in collecting are the same as those described for Ascomycotina (pp. 15–16) but because of the small size of most of them a keen eye is required. Some Hyphomycetes form distinct brown, black, green, pinkish, rose or whitish felt-like growths on leaves, decaying fruits, rotting wood and bark but many are relatively inconspicuous. Leaves which appear to be diseased are worth taking back to the laboratory for examination with a binocular microscope; no fungi will be found on many but small tufts of conidiophores arising through stomata or associated with the margins of leaf spots will sometimes be detected in this way. Others may be stimulated to develop by placing them in damp chambers (see p. 18).

Myxomycota

Myxomycetes are often abundant in woods two or three days after heavy rain. They are most commonly encountered on rotting wood but some also occur on dung and damp straw. Specimens should be removed together with the wood on which they are growing with the aid of a sheath knife and are best pinned to cork in small boxes or tins as they are collected to reduce the risk of damage to their often delicate fruiting structures. Plasmodia are often conspicuous and brightly coloured but difficult to name. If plasmodia are collected and placed near a window in the laboratory in a petri dish lined with moist filter paper, however, the fruiting structures often develop in a few days. The specimens should be dried slowly and glued into boxes for permanent storage. Members of the Labyrinthulales are isolated by placing algae on 1% serum-sea water agar when they appear after 2–7 days. For further information see Carlile (1971).

Tropical microfungi

The fungi of tropical regions are still very imperfectly known and many species remain undescribed. In wet tropical forests saprophytic and weakly parasitic fungi tend to be less host specific than in more temperate regions. Some tropical plants, however, have extremely rich and distinctive floras and the leaves of evergreen vascular plants are often covered with mosaics of fungi and lichens. Hyperparasites (i.e. fungi growing on other fungi) are also common, although these will be found on examination in the laboratory rather than in the field. As the vascular plant flora in some tropical areas is still relatively imperfectly known particular care is needed in naming hosts. For further information on collecting in the tropics see Deighton (1960).

Aquatic fungi

Many fungi occur both in fresh-water and sea-water on fish, algae, wood and other decaying materials. Phycomycetes often form fluffy growths around twigs and on fish in slow flowing streams, lakes and ponds, and many interesting Hyphomycetes are often associated with decaying leaf skeletons. Specimens are placed in polythene bags or tubes together with water from the site from which they have been collected and kept cool until required for examination. Collecting in this way, however, does not lead to a comprehensive knowledge of the fungi present in such habitats as many are opportunist and wait for the correct substrate to drop into the water. Baiting techniques, in which material (e.g.

various seeds, fruits, bread, meat, twigs) is placed in small wire mesh baskets and suspended in water by nylon cord and examined after 1–5 weeks, have proved particularly valuable. Some chytrids and ascomycetes are commonly encountered on freshwater and littoral algae, respectively. A comprehensive account of techniques for collecting and isolating aquatic and marine fungi is provided by Jones (1971).

Soil fungi

Soils have extremely rich fungal floras in addition to the larger basidiomycetes which produce their sporocarps on the surface of the soil. Soil mycofloras can be investigated by mounting soil on a slide and examining it directly or by a variety of isolation techniques (see Burges, 1958; Parkinson and Williams, 1961; Barron, 1971). Hypogeous macrofungi (e.g. truffles) are frequently overlooked in floristic accounts. These usually occur in the upper layers of leaf litter and soil in woodlands and are traced by using a small rake to remove the litter, humus and soil to a depth of a few inches. The depth at which hypogeous fungi occur and what they are attached to should be noted. Once discovered they are treated as described for fleshy fungi (pp. 14–15) but they should also be cut open and careful notes made of their internal tissue arrangements. A useful summary of hypogeous fungi is provided by Hawker (1955).

Coprophilous (dung) fungi

Many members of the Mucorales and some ascomycetes, basidiomycetes and myxomycetes are known only from the dung of various animals and birds. Dung at different stages of decomposition has characteristically different mycofloras and so many dung types in various states should be studied. If pieces of dung from different animals are collected and placed in 'damp chambers' (petri dishes lined with moist filter paper; see Dade and Gunnell, 1969) the succession of species can be followed and many fungi not detected in the field can be found in this way. Lundqvist (1972) provides a useful account of methods for detecting coprophilous pyrenomycetes. Dung can be dried, fumigated, and kept in herbaria in paper packets or glued into small boxes.

Airborne fungi

Airborne fungal spores can be trapped by exposing glycerine jelly or vaseline coated slides, or petri dishes to the air for varying lengths of time, or by more complex quantitative sampling techniques (see Davies, 1971; Ingold, 1971). A few fungi can be named directly from their spores but many require subsequent culturing to establish their identities with any great degree of certainty.

Fungi on man and animals

Hairs, skin and nail scrapings, feathers and horns infected by dermatophytes can also be collected. These may retain their pathogenicity for several years but are rendered harmless by fumigation in petri dishes between layers of filter paper soaked in formalin or by treatment with propylene oxide. They may then be placed in transparent paper packets or glued into small boxes. Laboulbeniomycetes are often most satisfactorily preserved in 70% alcohol. A detailed account of techniques used in the collection and isolation of many fungi which occur on man and animals is provided by Stockdale (1971).

EXAMINATION

WHEN the collector returns to the laboratory he must begin to examine his material. Examination starts with a study of the macroscopic features visible to the unaided eye, hand lens ($\times$ 10–20) and dissecting microscope ($\times c.$ 20–80), and proceeds to microscopic ($\times c.$ 150–1000) and microchemical characters. Where the material seems to represent something which the collector has not seen before it is helpful to make sketches, notes and camera-lucida drawings and to take photographs as he proceeds. All notes, drawings, photographs and microscopic preparations relevant to particular collections may either be placed with them in the herbarium or retained in a series of loose-leaf folders. There are no adequate descriptions or drawings of many microfungi and the collector will find that his own notes are valuable for reference when he comes to name other collections in the future.

If, after a collection has been examined, it proves to be immature or too poor for certain determination it is best to discard it so as to prevent it becoming a hindrance in the future. All specimens which are eventually destined for a herbarium are allotted collector's numbers. Many mycologists and lichenologists employ a simple numerical sequence throughout their life (e.g. 1–35,000) but others use different series of numbers for each expedition (e.g. K 1126 for Kenyan expedition collection no. 1126) or for collections made in each year (e.g. 72/1126 for collection no. 1126 made in 1972). A series of notebooks listing numbers and details of each collection may be kept but if for any reason a collector is uncertain as to the number last used it is best to skip several tens or hundreds of units to avoid the possibility of duplication. Numbers are useful for labelling any drawings, slides or photographs prepared and are also of considerable value when sending parts of collections to specialists (see p. 49). Their main purpose, however, is to enable particular collections to be referred to accurately in publications.

Macroscopic examination

On returning from a collecting trip where material from different groups has been collected the mycologist must first sort the specimens by eye into their main groups. Those which will deteriorate most rapidly (e.g. Agaricales) or ones likely to be grown in culture require priority while those which will be little affected by leaving for some time (e.g. lichens) may simply be dried and kept for examination at a later date.

A useful account of the procedure to be followed in examining a member of the Agaricales is provided by Henderson *et al.* (1969). When a coloured sketch has been prepared, and the size, shape, consistency, odour etc. (see p. 15) noted, the pileus (cap) is cut off and placed gills-downwards on a piece of either black or white paper (ideally across the boundary of a half-black and half-white card) and left for 15 minutes to a few hours to obtain a spore print. Better prints may sometimes be obtained by placing a drop of water on the top of small caps and placing a box or other container over them to reduce evaporation. The colour of spores is easily seen from spore prints and they also often provide clear impressions of the gill patterns. Some method of standardizing spore prints is essential in taxonomic work employing this character and a way of doing this has been suggested by Watling (1970). Although water-coloured drawings are the best way to indicate colours time may not permit these to be prepared. Some mycologists place dabs of the appropriate colour on their sketches while others employ a system using uncoloured sketches and a rectangle on each divided into blocks each of which is painted the colour of a particular part (e.g. upper block = pileus, centre block = gills, lower block = stipe). The value of colour photography in the fleshy fungi has already been referred to (p. 15).

Appropriate reagent tests on fleshy fungi should be carried out at this stage if they have not been made in the field.

Many macrolichens (i.e. foliose and fruticose species) can be named from macroscopic characters alone with the aid of a hand lens and a few chemical reagent tests, although in some cases microscopic and chromatographic examination may be necessary for confirmation. In examining macrolichens particular attention should be paid to the branching system, habit and size and to the location and form of any ascocarps, cyphellae, isidia, pseudocyphellae, rhizinae or soralia which may be present. Any colour reactions which occur on application of the standard reagents (i.e. K, C, KC and PD*) to either the thallus or medulla should also be noted at this stage. When testing the medulla the cortex has to be cut or scraped away with a razor, and in the case of tests performed on dark coloured specimens any colours produced are more readily seen if absorbed on to soft tissue or filter paper. The portion tested should always be discarded.

In studying most microfungi and many of the smaller (i.e. crustose) lichens relatively little information can be obtained without the use of a dissecting microscope. In examining material with a dissecting microscope direct illumination from above the specimen is required. Particular attention should be paid to the arrangement of any sporocarps or conidiophores on the thallus, other stromatic tissue, substrate or host, noting whether they are single, aggregated, immersed, erumpent, sessile or stalked. If discocarps are present any characteristic margins (e.g. a thalline margin in lichenized species, hairs in the Pezizales and Helotiales) should be noted. In pyrenomycetes the ostioles often require careful study and any associated imperfect state should be looked for. In the case of Coelomycetes the distinction between acervuli, pycnidia and stromata is particularly important, and in the Hyphomycetes coremia, synnemata or sporodochia should be searched for and whether the spores are produced dry or in slimy masses noted.

All gross structures which may be of value in identification can be measured

* K = 15 % aqueous solution of potassium hydroxide; C = bleaching powder or a commercial bleach preparation (e.g. 'Parazone'); KC = application of K followed by C; PD = freshly prepared solution of p-phenylenediamine (5 %) in alcohol, or Steiner's solution (1 g p-phenylenediamine, 10 g sodium sulphite, 100 ml water, c. 10 drops of a liquid detergent) which is stable for at least six months. PD is a toxic

at this stage using either an ocular micrometer or a rule graduated in units of 0·1–0·5 mm.

The dissecting microscope is also useful for removing samples for microscopic examination and spores or other inoculum for the preparation of cultures.

Microscopic examination

An adequate high-power microscope providing magnifications of up to ×1000 is essential for examining both larger Hymenomycetes and minute Hyphomycetes alike. Information as to the types of microscope required, their setting up and use is provided by Laundon (1968). For routine examination a minute portion of the sporing tissue is removed under a dissecting microscope with the aid of a mounted needle, needle-knife, razor blade or scalpel point, and placed in a small drop of mounting fluid on a microscope slide. If bulky the material is teased apart gently and a coverslip applied. Some light pressure on the coverslip (e.g. by tapping with the butt of a pencil) serves to disperse the structures in a thin optical plane. Warming the slide over a spirit lamp or small bunsen flame, or placing it on a warm microscope lamp case will cause any air bubbles that may be present to disappear but the mounting medium should not be allowed to boil.

The most satisfactory general purpose mounting medium for fungi is Amann's lactophenol including a stain such as cotton blue. This medium has the advantage that the slides can then be readily preserved by sealing them with nail varnish or other suitable proprietary sealant but is not suitable for preparations of lichens growing on calcareous substrates. If a slide in lactophenol is to be preserved it must be warmed for an hour or so to ensure that air bubbles are completely eliminated and any excess mounting fluid throughly cleaned off. Two layers of nail varnish are then applied, a clear one first and then a coloured layer when the first has dried. This two-coloured nail varnish procedure has the advantage that it is always readily apparent whether one or two layers have been applied. Slides preserved in this way are reasonably permanent and may be kept with the material from which they were made in specially constructed slide-boxes. Tribe (1972) recommends the use of 'Glyceel' instead of nail varnish, and Laundon (1971) that a layer of PVA (polyvinyl acetate) water-based gluc bc placed over a layer of nail varnish to make the slide more durable.

Iodine (most frequently used as Melzer's Reagent) is widely used in the study of both lichenized and non-lichenized Ascomycotina and some macromycetes. Eriksson (1970) provides details of Minks' technique for obtaining the most satisfactory results for iodine ('amyloid') reactions of apical apparatus in asci. Water is commonly employed as a mounting medium by some mycologists but is not satisfactory for most purposes as the slides dry out very rapidly and, as is the case with Melzer's Reagent, are difficult to preserve adequately. In examining both tough conglutinate ascocarps (such as those which occur in the lichen-forming Lecanorales) and hymenial tissues of macromycetes, a 10–15% solution of potassium hydroxide in water is very satisfactory; slides made with

compound and may also discolour herbarium paper and specimens (see Hawksworth, 1971a).

this reagent should only be left on the microscope stage for the minimum amount of time as some corrosion of the objectives may occur. Hyaline epispores (exospores) and appendages of spores (such as those that occur in some genera of the Sordariaceae) are most readily seen if mounted in black Indian ink (see Lundqvist, 1972). Conidiogenous structures and septa in hyaline Deuteromycotina are shown up clearly by mounting in a solution of 0·5 g erythrosin in 100 ml of 10% ammonia. Dematiaceous Hyphomycetes may have some of their structure obscured by using stains with cotton blue and may simply be mounted in pure Amann's lactophenol. For detailed accounts of the composition of mounting media and stains for mycological material see Ainsworth (1971), Commonwealth Mycological Institute (1968) and Dring (1971).

A valuable method for studying the microfungi on leaf surfaces is the cellulose acetate 'Necol' mounting technique (Ellis, 1950; Dring, 1971). An alternative to this method is to use adhesive transparent tape (e.g. Sellotape) to strip off the fungus. In this latter case drops of lactophenol can be placed above and below the tape on a microscope slide and a coverslip applied.

To ascertain the structure and arrangement of sporocarps accurately, vertical sections will sometimes be required. Although it is sometimes possible to prepare adequate sections by hand using a single-edged razor blade or 'cut-throat' razor and holding the material between some tissue such as elder pith or under the dissecting microscope, a freezing microtome provides the most satisfactory results. Freezing microtomes form an essential part of a mycological laboratory's equipment and consist of a fixed stage cooled to below the freezing point of water (either by carbon dioxide or a 'Pelcool' cooled water system) and a movable blade calibrated to cut material frozen on to the stage with a drop of dilute gum arabic in water at a range of thicknesses; 5, 10, 15 or 20 μm (10 μm sections are the most commonly used at the CMI). The sections are collected with a fine paint brush as they are cut and placed directly in a drop of mounting fluid on a slide or in water in a watch glass. When dealing with particularly hard material such as hardwood blocks sledge microtomes (freezing or not) are necessary. Dried specimens to be cut by freezing microtomes often require soaking overnight in gum arabic to soften them prior to sectioning. Embedding in paraffin wax is suitable for particularly delicate tissues or material to be sectioned on a non-freezing sledge microtome but this procedure is too time consuming for routine use when large numbers of specimens have to be examined. Dring (1971) provides further information on embedding and microtome techniques.

Microscopic examination by normal transmitted light is adequate for most purposes but dark-field and phase contrast sometimes prove of value when studying indistinct (e.g. spore ornamentation) or hyaline structures. Particular attention should always be paid to the way in which the spores are produced (e.g. nature of conidiogenous cells, asci, basidia), the arrangement of the spore producing tissues in any sporocarp that may be present, any characteristic sterile tissues amongst or surrounding the spore producing cells (e.g. cilia, nature of any inter-ascal tissues, cystidia), and to the structure and arrangement of the spores. Accurate measurement of microscopic structures which will be needed to identify the material are essential.

Microscopic measurements are made with a micrometer eyepiece but before the eyepiece scale can be used it must first be calibrated. Each microscope must be calibrated individually and this is achieved by placing a graduated slide (i.e. a slide with a scale accurately divided into divisions each of which is, for example, 10 μm) on the microscope stage and measuring the number of divisions on the eyepiece scale which correspond to one or more divisions on the graduated slide. From this information the distance each division of the eyepiece scale represents can be very easily calculated. A separate set of calibrations is required for each eyepiece and objective used and many mycologists draw up a table to show what each eyepiece division (and multiples of it) represent with particular objectives. Using such a table (Table 1) the correct measurements may be read directly from it without constantly having to make mental calculations. In the course of routine examination work the largest and smallest mature spores should be searched for and measured to ascertain the range of variation which occurs; between five and ten individual spore measurements are usually necessary.

Table 1. *Calibration chart for a microscope eyepiece micrometer*

Objective		Number of units on the eyepiece scale and the distances they represent (in μm)								
		1	2	3	4	5	6	7	8	9
Without drawing tube	13/1	6·9	13·8	20·7	27·6	34·5	41·4	48·3	55·2	62·1
	25/0·5	3·7	7·4	11·1	14·8	18·5	22·2	25·9	29·6	33·3
	63/1	1·5	3·0	4·5	6·0	7·5	9·0	10·5	12·0	13·5
	100/1·3	0·9	1·8	2·7	3·6	4·5	5·4	6·3	7·2	8·1
With drawing tube	13/1	5·3	10·6	15·9	21·2	26·5	31·8	37·1	42·4	47·7
	25/0·5	2·8	5·6	8·4	11·2	14·0	16·8	19·6	22·4`	25·2
	63/1	1·1	2·2	3·3	4·4	5·5	6·6	7·7	8·8	9·9
	100/1·3	0·7	1·4	2·1	2·8	3·5	4·2	4·9	5·6	6·3

Measurements need only be given to the nearest 0·5 μm but the lengths and breadths and sizes of any appendages all need to be ascertained. Measurements normally include any epispore or ornamentation but where these are conspicuous these should also be measured independently. The sizes of larger structures such as globose sporocarps (e.g. perithecia, pycnidia) are difficult to ascertain from squash preparations and measurements of these structures may most satisfactorily be determined from unsquashed or microtome sectioned material.

Useful accounts of methods and procedures for microscopically examining members of the Agaricales are included in Henderson *et al.* (1969) and Zoberi (1972); Swinscow (1962), Duncan (1970) and Harris (1973) provide details of microscopic techniques used for studying lichens.

If the microscope being used has a photographic tube designed to take a 35 mm camera it is convenient to have a loaded camera permanently attached so that any interesting structures found may be photographed as examination proceeds. Similarly, many mycologists prefer to keep a camera-lucida device (see p. 59) fixed to their microscopes so that accurate drawings can be made as required.

Microchemical examination

The determination of the chemical components in lichens forms an important part of the examination of some genera (particularly foliose and fruticose species). A few fragments of the lichen are crumbled on to a microscope slide either on a slide warming plate or at room temperature and acetone added drop by drop from a pipette. After a few minutes a ring or residue of crystals forms around the fragments and when the slide is dry the fragments are brushed off with a fine paint brush. The compounds present in the residue can be determined either by microcrystal tests (MCT) or thin-layer chromatography (TLC).

To employ microcrystal tests effectively it is necessary to know what compound or compounds are being tested for but by chromatography most of the characteristic substances which are used as taxonomic characters can be identified easily (and ones likely to have been overlooked by microcrystal tests discovered). Sufficient residue is often obtained from a single acetone slide extraction to enable both microcrystal tests and thin-layer chromatograms to be made from it.

To perform microcrystal tests a coverslip or portion of a coverslip is placed over a whole or part (if it is being used for other microcrystal tests or chromatography) of the residue. The minimum amount of one of several crystallizing agents* is then added, the slide warmed over a spirit lamp until the residue has dissolved (taking care not to allow the reagent to boil), allowed to cool for 15–30 minutes, and examined microscopically by polarized and non-polarized light (the latter is adequate for most substances). There is no single account of the crystal forms of all lichen products known to give microcrystal tests and the works of Hale (1967), Thomson (1968) and papers cited in Hawksworth (1971a) should be consulted. The most satisfactory way to become familiar with the form of crystals of particular compounds and their variability in particular reagents is to study species whose chemical components are well known (see C. F. Culberson, 1969, 1970; Dahl and Krog, 1973).

For thin-layer chromatography a drop of acetone is added to the residue and then mixed with and taken up into a disposable capillary pipette or tube and spotted on to a plastic-coated sheet (e.g. Eastman Chromagram sheet no. 6060), an aluminium-coated sheet (e.g. Merck Silica Gel F-254) or a glass-coated plate. The most widely used solvent is toluene(90):dioxane(25):acetic acid(4). When the solvent front has advanced to 7–10 cm (depending on the size of the plate or sheet used) the chromatogram is dried, observed under ultra-violet light (the colours of spots noted), and sprayed with filtered Steiner's solution (see p. 20) or 10% sulphuric acid (with heating; aluminium sheets and glass plates only). When running chromatograms controls (i.e. residues from species of known chemistry) should always be employed. For identification of chromatogram spots see Culberson and Kristinsson (1970), Culberson (1972) and Santesson (1967).

It should be emphasized that some of the chemicals employed in carrying out

---

* The most useful crystallizing reagents are GE, glycerine (1):acetic acid (1); GAW, glycerine (1):alcohol (1):water (1); GAoT, glycerine (2):alcohol (2):o-toluidine (1); and KK, 5% aqueous potassium hydroxide (1):20% potassium carbonate (1).

reagent (see p. 20), microcrystal and chromatographic examinations of lichens are harmful and should be handled with caution (see Hawksworth, 1971a).

HERBARIA

HERBARIA (sing. herbarium) are places where dead (usually dried) material of plants is permanently preserved. The larger mycological herbaria contain many tens of thousands of specimens including type material (see pp. 179–192) and other important collections. Such herbaria are invaluable to systematists as they serve as reference collections and contain material which needs to be examined in the course of revisionary and monographic studies. Most newcomers to systematic mycology will not have access to one of the major mycological herbaria and will find that they need to build up their own reference collections of dried specimens. If such a personal herbarium later comes to include specimens cited in published papers care should be taken to ensure that it will ultimately be deposited in one of the major national herbaria, at least after the mycologist's death. Many important collections have been lost, destroyed or mislaid simply because their value was not realized at the appropriate time.

In some larger herbaria mycological specimens constitute a relatively small part of the total collections and the methods by which they are handled and arranged tend to be the same as for other plant groups. The CMI herbarium has, however, been developed solely for fungal specimens and the following sections outline some of the techniques and methods used there. More detailed accounts of the ways in which material at CMI is handled are provided by Ellis (1960) and Onions (1971).

Specimens

A herbarium specimen may be a single sporocarp or a portion of one (e.g. in Aphyllophorales and Agaricales), dried culture, slide, or the material on its host or substrate (e.g. leaf, stem, bark, rock, soil, paper, cloth). Most microfungi and lichens are adequately preserved simply by drying (see pp. 15–16) and by the time they reach the herbarium this process should have been completed. Crustose lichens on soil present a particular problem (p. 16) but if the soil is bound by painting it with substances such as 'Rhoplex AC-33' (Gates, 1958) or 'Primal AC-55' (Wade, 1959) diluted with water and allowed to dry thoroughly overnight adequate herbarium specimens may be made.

At one time botanists endeavoured to collect and preserve a single specimen of each species and discarded the one they had in preference to any larger specimen they came across regardless of the locality. Today mycologists keep as much material as is practicable of each species. A large number of collections is often essential to provide a clear concept of the full range of variation of a species, its geographical distribution, and to provide adequate material for study by future workers. Any specimens used in studies which are published need to be deposited in herbaria. In addition to material cited in floristic and taxonomic studies that employed in chemotaxonomical, biochemical, physiological, serological, pathological or genetic studies should be included. If such material is not preserved it may not be possible to be certain what species an author used in the future as concepts of particular taxa may be altered over the years (as a result of more detailed taxonomic investigations) and any queries can be readily checked. *Polyporus tomentosus*, for example, was reported to be a source of galactose oxidase but later workers were unable to confirm this; the original material had fortunately been preserved and it was discovered that the original report was based on *Dactylium dendroides* which was parasitic on *P. tomentosus*. Where the material is to be destroyed in the course of a physiological or chemical study at least a portion should first be set aside for the specific purpose of being a voucher herbarium specimen. Ideally fungi used in experimental studies should also be preserved in the living state (pp. 35–37).

Simply because of lack of space it is sometimes impossible for herbaria to keep all the material of common species they are sent but before any material is discarded care should be taken to ensure that it is not from a new geographical area, has not been and is unlikely to be referred to in a published report, and that it comes within the range of variation of the species already represented by adequate material in the herbarium.

Slides

Any slides made from a specimen are kept together with the specimen from which they are made at the CMI. This procedure facilitates any subsequent examination and prevents additional parts of the specimen having to be removed each time anyone wishes to check it, a consideration which is particularly important in the case of type specimens. In some cases removal of fragments from a particular collection for microscopic examination by different workers over the years has rendered them almost useless today. At the CMI slides made with lactophenol are preserved by ringing with nail varnish or glyceel (p. 21) and placed in specially made card slide boxes which are kept with the specimen itself. Some herbaria maintain separate slide collections in wooden cabinets with shallow drawers made specially to contain slides and affix labels to herbarium specimens to indicate that a slide is in existence but housed separately in their 'slide collection'.

Cultures

Cultures now constitute an extremely important part of mycological studies and dried cultures also have a place in the herbarium. To prepare dried cultures 1·5% tap water agar and glycerol is melted and poured on to the lid of a plastic Petri dish; the culture is removed from the other half of the Petri dish and placed in the centre of the lid or on a piece of hardboard. In the case of tube cultures these are removed from their tubes with a scalpel and laid on molten tap water agar on a smooth piece of hardboard to which an adhesive label bearing the number of the culture has also been attached. The cultures are then allowed to dry in a drying cupboard (to prevent the dispersal of spores into the laboratory) at room temperature for 12–72 hours (depending on the thickness of the agar and the humidity of the air). When the agar is completely dry a razor blade is used to cut through the tap water agar (for material on hardboard) or the culture is eased out of its Petri dish with a scalpel. The dried plate cultures are then glued to a cardboard ring and fixed in a protective cardboard box (Fig. 1); while the dried tube cultures are glued to the base of slide boxes. At the CMI the dried cultures are placed in the herbarium packets together with any slides made from them and the original material from which the cultures were made (if this is available). Onions (1971) provides an illustrated account of the steps involved in preparing dried cultures for the herbarium.

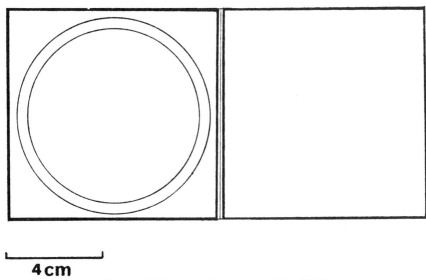

4 cm

Fig. 1. Dried culture box as used at the CMI.

The maintenance of living cultures is discussed in the following section of this book (pp. 35–37).

Packets

All material of a particular collection is most appropriately kept in a packet of a standard size made from good quality stiff paper. At the CMI packets 15 × 10·5 cm (6 × 4 in) when folded are used (Fig. 2). Packets of this pattern are the most suitable for sticking to herbarium sheets as all four flaps may be opened to expose the contents. Very large specimens are kept separately in boxes and notes placed on appropriate sheets in the herbarium to indicate their existence. Specimens to be placed in herbarium packets at the CMI are first inserted into transparent paper envelopes. Fragile specimens may be wrapped in soft tissue paper and lichens are often best glued to 5 × 3 in blank card-index cards. Protective boxes (c. 5 × 5 × 0·5 cm) glued to cards or herbarium sheets are also useful for preserving material likely to be damaged by friction. All items within a particular packet should be clearly numbered individually with the herbarium accession number and (or) collector's number.

The type of packet used at the CMI is not, however, suitable for material to be stored in drawers or boxes as they tend to fall open. An alternative type of

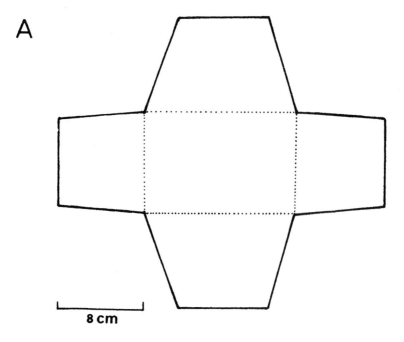

8 cm

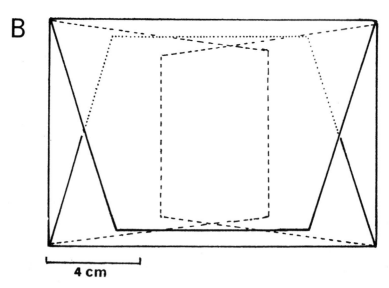

4 cm

Fig. 2. Herbarium packet suitable for attachment to herbarium sheets. A, Unfolded; B, folded.

packet suited to storage in this way may easily be made from a sheet of 22·5 × 20·5 cm (10 × 8 in) sheet of good quality white paper (Fig. 3). In some cryptogamic herbaria this type of paper packet is then placed within an ungummed stiff brown paper envelope bearing the label to ensure that it remains firmly closed.

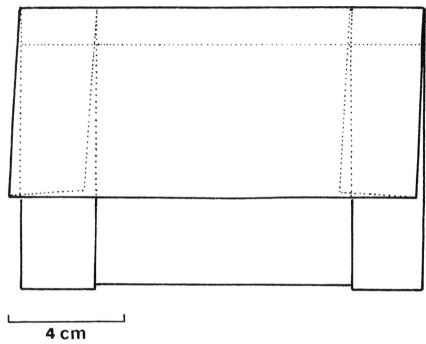

4 cm

Fig. 3. Herbarium packet prepared by folding a piece of paper suitable for storage in drawers or boxes.

Labelling

At the CMI standard labels 12·6 × 7·6 cm (5 × 3 in) printed in black are glued to each packet and carry all pertinent collection data (Fig. 4). Each packet should ideally bear the following information: (1) collector's name and number, (2) date of collection or isolation, (3) locality of collection (i.e. country, province, county, parish, locality name and grid reference or latitude and longitude), (4) altitude (for lichens), (5) the substrate or host* from which it was collected or isolated (given as precisely as possible), (6) any hands through which it passed (e.g. material received in exchange or via other herbaria or culture collections), (7) the name of the person who identified it and the date it was identified, (8) a reference to any correspondence filed elsewhere concerning the material, and (9) a note of any published papers in which the particular material was cited. The herbarium accession number is placed on a self-adhesive label and attached to the top left-hand corner of the packet, and a label in red is placed on the top right-hand corner of collections which are type material.

* Hosts should always be referred to by their currently accepted scientific names. For scientific names of economic plants see Purseglove (1968, 1972) and Uphof (1968).

The information put on the labels should be either typed or legibly written in an ink which will not fade. When many collections have been made in a particular locality it is often convenient to have printed labels bearing this information prepared.

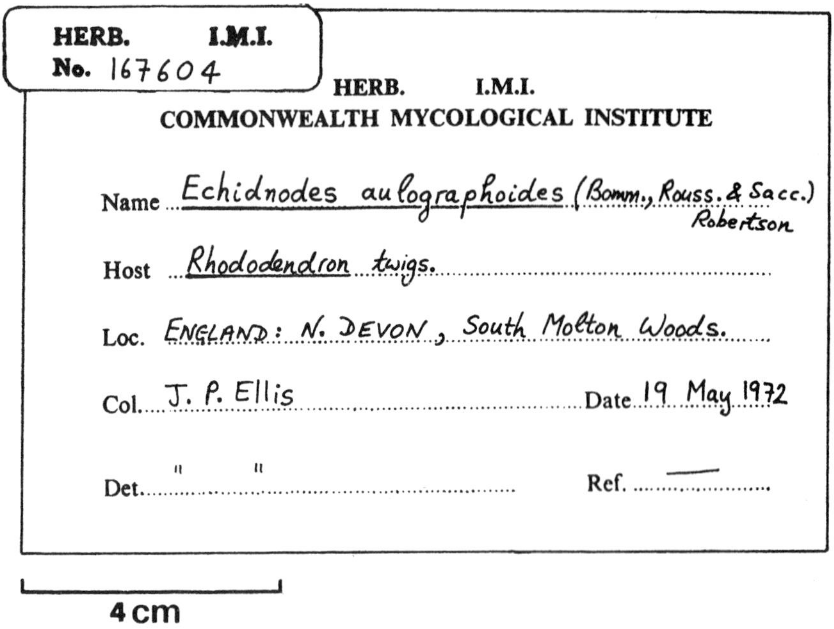

Fig. 4. A CMI herbarium packet label.

Duplicates

When a collection is large it is often possible to divide it into a number of duplicate packets. The label on each duplicate packet bears the same information and number as the packet from which it was taken. Duplicate collections in the CMI are exchanged for named collections of microfungi from other main mycological herbaria from time to time but beginners will also find that an exchange of duplicates between themselves provides a means of broadening the scope of their personal collections.

Herbarium records

Each specimen that arrives at the CMI is given a herbarium accession number which is unique to that particular collection. Particulars of each collection are entered next to the serial accession number in a strongly bound ledger referred to as an 'accession book'. The serial number appears at the extreme left of the page, the country of origin and collector in the adjacent column, and details of the collection (i.e. host and locality) on the remaining portion of the page. These data are all included on a left-hand page and the adjacent right-hand page is left blank. When the material has been named or checked by a mycologist this name is entered on the right-hand page as it is under this name that it will be placed in the herbarium. If for any reason a particular specimen's name is changed at any time in the future this is also indicated on the right-hand page. Using this method it is consequently possible to trace the whereabouts of any of the over 170,000 specimens which have passed through the CMI's accession books however many times its name may have been changed as long as its accession number is known.

Two other series of record books are maintained at the CMI: (1) a 'fungus book' which lists all the collections in the herbarium of a particular fungus (with one or more pages to each fungus), and (2) a 'host book' which provides a list of all the fungi represented on particular hosts at the CMI. These series are maintained in loose-leaf books so that additional pages can be intercalated as required. By employing the host and fungus books it is possible to extract items of information on particular hosts or fungi readily without recourse to examining the herbarium packets themselves.

The maintenance of herbarium records in this way is very time consuming and in some large herbaria steps are now being taken to mechanize this aspect of herbarium work by employing computers. For accounts of the value and application of data-processing techniques to herbarium collections see Argus and Sheard (1972), Soper (1969), Hall (1972a, b) and Krauss (1973).

Fumigation

Before specimens are mounted and put away in herbarium cupboards at the CMI they are fumigated by exposure to methyl bromide in a fumigation chamber for 24–48 hours to prevent contamination and damage by mites or other insects. Periodic fumigation of folders in herbarium cabinets is also carried out from time to time and while this is being done the cabinets themselves have been sprayed with 'Lindex'. Dichlorvos and pyridine have also been used as fumigants in some herbaria. Naphthalene and paradichlorobenzene ('moth balls') are often placed in herbarium cabinets to prevent damage by insects in some herbaria. The use of lindane in herbaria, however, is not now recommended because of its toxicity and dichlorvos has to be handled with care (see Lellinger, 1972).

Mounting and incorporation

Herbarium packets are glued with gum arabic (so that they can be removed easily if this becomes necessary) on to herbarium sheets of thick white paper $45 \times 25 \cdot 5$ cm ($16 \cdot 5 \times 10$ in). Specimens of the same species from different hosts (for fungi) or countries (for lichens) are attached to separate sheets. Six to eight packets may be placed on each herbarium sheet. All sheets of a particular species are placed within a white paper folder, and all species folders within a brown card genus folder. The name of the genus is placed on the brown folder, and species names on the bottom of the species folders. Each sheet is labelled with the name of the host or country from which it bears specimens. If the type specimen of a particular species (or its synonyms) is present this is placed on a separate sheet at the front of the species folder. Type specimens require careful curation because they are irreplaceable. In some major herbaria type specimens are placed in separate folders edged in red to both safeguard them and draw attention to the fact that they are types and should always be treated with considerable care. Type specimens should not be damaged or have parts removed for the purposes of routine identification work (as opposed to taxonomic research) and to prevent this some herbaria place all their type specimens together in one part of the herbarium away from the general collections and keep them in locked cabinets.

The mounted material is housed in specially constructed wooden cabinets. If wooden cabinets are employed it is essential that they are well dried and treated with insecticide before they are used. Metal cabinets are now being used instead of wooden ones in many large herbaria. The herbarium cabinets themselves must be kept in a cool and dry atmosphere to reduce the risk of invasion by unwanted moulds. In tropical countries ventilation and periodic drying and airing of the cupboards is necessary if an air conditioned room cannot be provided. Many valuable hints on the maintenance of herbaria in tropical climates are included in Fosberg and Sachet (1965).

If packets are to be kept individually and not attached to herbarium sheets these may be filed in order in wooden or metal filing cabinet drawers of an appropriate size. Most personal herbaria start as filed systems in cardboard boxes (e.g. shoe boxes) but as more and more specimens are accumulated this method of storage begins to prove unsatisfactory.

The way in which genera are arranged in the herbarium is largely a matter for personal preference but the method used must enable particular species to be located with the minimum of difficulty. At the CMI the genera within each major group of the fungi are arranged alphabetically and the species within each genus are also placed in alphabetical order. Some exceptions are made in the case of the Uredinales and some genera of Coelomycetes (e.g. *Phyllosticta*) when the fungi are arranged according to the families of their host plants. Within the larger orders of Ascomycotina the genera are placed in alphabetical order within Saccardo's spore groups.

Various other systems for the arrangement of mycological herbaria have been proposed (see Bartholomew, 1931; Jefferys, 1972); some of the older national herbaria employ the numbered systems of genera and species adopted by Saccardo (1882–1931) and Zahlbruckner (1921–40). For most purposes, however, simple alphabetical systems are most satisfactory as these allow additional genera and species to be intercalated most easily. Where all genera are not placed in a simple alphabetical system a 'deposition book' is required in which the location of all genera represented in the herbarium is indicated.

Loans

Any herbaria which contain material that has been cited in published work will from time to time be asked to send material on loan. Considerable caution needs to be exercised in lending specimens, however, as many important collections have been lost in the post or never returned by the borrower. To reduce the risk of losing irreplaceable collections many herbaria are now careful only to send material to recognized herbaria or to specialists whom they know will treat it with care and return it promptly and in good condition. When writing for the loan of specimens the prospective borrower should indicate why he requires the particular collections. Where particular collections are large the CMI usually sends only a portion of the material on loan after care has been taken to ensure that the piece sent is representative. Some important early collections (e.g. those of Acharius and Linnaeus in H* and LINN*, respectively) are not sent on loan at all.

Careful records should always be kept of material sent out on loan and notes placed in the herbarium to indicate where collections which are missing have been sent. Most herbaria send forms to borrowers to sign and return on receipt of the specimens and loan material for maximum periods (usually three, six, or twelve months). Material should always be returned as soon as it is finished with, however, and if for any reason an extension of the period of a loan is required this should be applied for before the term of the loan expires.

Any specimens sent through the post should, of course, be carefully packed to reduce the risk of damage in transit. Some herbaria require that their material should be sent only by registered or insured postal services.

CULTURING

THE study of material in artificial culture now constitutes an important aspect of systematic studies in many groups of fungi. Only by growing material on the same medium under comparable conditions can assessments of the extent of the effects of the host or substrate on morphological characters be made. Comprehensive reviews of techniques employed for the isolation and culturing of fungi are included in Booth (1971a) and a brief introduction to them is provided in Dade and Gunnell (1969). This section is intended only to serve as an introduction to the available methods paying particular attention to those which are most used at the CMI and will prove of value to those embarking on cultural studies with fungi for the first time.

Isolation

The isolation of plant pathogenic fungi is often complicated by the occurrence of secondary superficial saprophytic species. To remove these the infected material first has to be 'surface sterilized'. Surface sterilization may be effected by a variety of procedures (see Booth, 1971b). Prolonged washing in water together with some form of agitation will remove many of the saprophytes but almost all may be definitely killed by placing a portion of the infected tissue in a screw-cap jar into which 2 ml of propylene oxide is poured over a small cotton wool pad. After about 30 minutes with the cap screwed down the material is removed, thin slices cut from it, and the slices placed on suitable nutrient agar.

Single spore isolates (i.e. cultures derived from a single spore of the fungus) are invaluable to taxonomists endeavouring to establish relationships between perfect and imperfect states of fungi. There are many methods of making single spore isolations (see Booth, 1971b). Single spores can be obtained by preparing

* See pp. 187–192 for herbarium abbreviations.

a spore suspension on a microscope slide and removing individual spores with a capillary pipette; by removing individual spores from an agar disk using a special cutter mounted on a microscope's objective; or by picking off single spores free-hand with the aid of a microscope.

References to papers providing detailed accounts of methods for isolating fungi from some specialized habitats are included in Booth (1971a) and on pp. 17–19.

Culture work has not yet been widely employed in studies of lichen systematics but taxonomists should remember that the fungal and algal components of lichens may be grown in artificial culture (Ahmadjian, 1967a; Richardson, 1971) and that it is becoming possible to grow intact thalli in culture and under controlled environmental conditions also (Ahmadjian and Heikkilä, 1970; Galun et al., 1972; Kershaw and Millbank, 1969, 1970; Dibben, 1971; Pearson, 1970). The use of transplant experiments in lichen taxonomy is discussed on p. 105.

Dates on which isolations are made should always be recorded as should the length of time particular cultures have been grown prior to drying down (see p. 26) or when maintained cultures were started or subcultured. Careful labelling and numbering of all plates and tubes is always essential to avoid the possibility of confusion as to the origin and date of starting of any particular culture.

Culturing

Many fungi have their own specific requirements for optimal growth in artificial culture and the most appropriate conditions for a particular fungus can be determined only by a series of tests involving ranges of temperature, light, pH, media etc. At the CMI the most widely used medium is potato-dextrose agar (PDA). This is prepared by cutting scrubbed mature main crop potatoes into 12 mm cubes, weighing out 200 g, placing these in 1 litre of water and boiling until soft. The potatoes are then mashed, squeezed through a fine sieve, 20 g of agar added, and boiled until the agar is dissolved. Dextrose (20 g) is then added and the whole stirred until it dissolves. The agar is placed in glass medicine bottles and, with their caps loosened, placed in an autoclave or pressure cooker at 15 p.s.i. for 20 min. After cooling the bottles are removed and their caps tightened. When some of the medium is required a bottle is placed in a water bath or pressure cooker (with the caps loosened) to melt the agar which can then be poured into plates (i.e. Petri dishes) or used to make agar slants in tubes. Plastic disposable Petri dishes are used at the CMI as, unlike many glass plates, these allow near ultra-violet (UV) light to penetrate (see below). Plates are inoculated by using the tips of mounted needles or loops of nicrome or platinum wire sterilized by heating to redness in a bunsen flame, or by dipping in ethanol and igniting in a bunsen flame.

Special media may be required for particular genera and groups of fungi and detailed accounts of the composition and use of many of these are given by Booth (1971a, c). Sterilized seeds, wheat straw, lupin stems, dung, filter paper etc. are often added to the surface of cooling plates of tap water agar (TWA) or potato-carrot agar (PCA) and these added structures are then inoculated with the fungus.

Some comments on preparing and maintaining cultures in tropical countries are included in Commonwealth Mycological Institute (1968).

At the CMI the plates are kept at room temperature and exposed to a 12 hour dark and 12 hour near ultra-violet (UV) cycle of illumination. The UV is provided by 'black-light' fluorescent tubes (peak 3650 Å, range 3100–4100 Å). Full particulars of this technique and its value in inducing sporulation in some microfungi are provided by Booth (1971c) and Leach (1971).

Slide cultures

In normal squash preparations the arrangement of the conidiophores, and the way in which spores are produced (e.g. chains or heads) may be difficult to observe. The method of slide culturing permits fungi to be studied virtually *in situ* with as little disturbance as possible. A piece of a suitable nutrient agar (7×7×2 mm) is placed on a sterilized slide and the fungus inoculated at the centre of each side of the agar square. A sterile coverslip is placed on top of the agar and the slide is placed in a damp chamber (i.e. supported on a pair of glass rods in a Petri dish lined with damp filter paper) in an incubator (see Fig. 5).

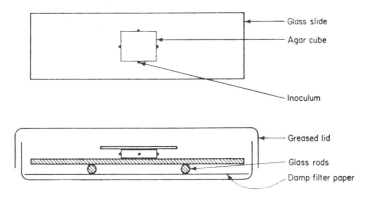

Fig. 5. Preparation of a slide culture (after Dade and Gunnell, 1969).

After about seven days the hyphae and conidiophores will be adhering both to the slide and coverslip. The coverslip is removed and placed colony growth down in a drop of a suitable mountant (e.g. lactophenol) on a clean slide. The agar square is also removed, a drop of mountant added to the growth adhering to the slide, and a clean coverslip applied. The slides can then be sealed using nail varnish (see p. 21). A detailed illustrated account of this technique is provided by Booth (1971b).

Culture collections

Just as a Botanic Garden is a living extension to a herbarium of vascular plants, a culture collection is a living extension to a mycological herbarium. Culture collections and herbaria are complementary to one another and not alternatives. The importance of preserving voucher specimens of material used in chemotaxonomical, biochemical, physiological, serological, pathological and genetical studies has already been emphasized (p. 25). Where the characters studied experimentally cannot be preserved in the dried state steps should be taken to ensure that the material is maintained in a culture collection so that future workers will be able to employ cultures derived from those used in original studies both to verify results and to extend the original observations. Where new taxa have been described, isolations made from holotype specimens or sub-cultures derived from the same isolation as a dried culture designated as a holotype should also be permanently preserved in the living state whenever possible. It should be emphasized here, however, that living subcultures them-selves cannot be nomenclatural types (see p. 127).

Before an institution starts to build up its own culture collection of fungi it is recommended that one of the established collections be visited to learn the methods employed. A list of the major mycological culture collections is included in Commonwealth Mycological Institute (1971) and Martin and Skerman (1972), and a detailed review of the techniques employed in these is provided by Onions (1971). Other pertinent information can be traced through Iizuka and Hasegawa (1970). The following methods of maintaining living cultures of fungi are the principal ones employed at the CMI.

(i) *Refrigeration.* Most microfungi will survive at 5–8°C and normal domestic refrigerators may consequently be used. If cultures are placed in refrigerated conditions their growth rates are reduced so that subculturing only needs to be carried out at six to eight month intervals. As some fungi may be killed by low temperatures a duplicate of each collection being kept by refrigeration should also be maintained by some other method. Domestic deep-freeze refrigerators are also suitable for maintaining fungi by refrigeration.

(ii) *Mineral oil.* Storage of fungi under mineral oil is a very cheap method of maintaining cultures and the lack of any need for elaborate and expensive equipment means that it is ideally suited to those entering this field for the first time. Cultures are prepared on slopes in glass or plastic McCartney bottles and covered to a depth of about 1 cm with a sterilized high quality mineral oil (e.g. medicinal paraffin oil of specific gravity 0·865–0·890). Mites, which may cause problems as contaminants in cultures (see p. 37), cannot penetrate the layer of oil. The screw caps on the tubes are left loosened and the tubes kept in trays in wooden cupboards. Cultures preserved in this way should be checked for viability at 2–3 year intervals and several generations (always including the oldest) main-tained. The value of this method of preservation is shown by the fact that of 2000 strains maintained by this method at the CMI for ten years only 45 were lost (Onions, 1971).

(iii) *Lyophilization (freeze-drying).* This method and the following one differ from the preceding two in that preservation is effected by suspending the metabolic activities of the fungus. The spores or cells of the fungus are suspended in a protective medium and small portions placed in glass ampoules and dried from the frozen state under vacuum. At the CMI a two-stage centrifugal freeze-dryer is now used. When freeze-drying is complete the ampoules are plugged and sealed under vacuum, stuck to filing cards, and kept in standard herbarium packets in filing cabinets. This method, although initially very expensive, has the advantage that many cultures can be kept in a relatively small space and that no contamination is possible as the ampoules are vacuum sealed. Also, because of the induced dormancy, cultural attenuation or mutation does not take place. At the CMI, however, only 60–70% of cultures have been found to be amenable to preservation in this way. A detailed account of the procedures involved in preserving living material of fungi in this way is provided by Onions (1971). The use of lyophilization in the preparation of herbarium specimens has already been referred to (p. 15).

(iv) *Liquid nitrogen storage*. In this method fungi are preserved by keeping them at the temperature of liquid nitrogen ($-196°C$) when their metabolism is at a standstill. This is a method of preservation suited only to the specialist laboratory which can justify the high expense of the costly apparatus and materials involved. Details of this technique are included in Onions (1971).

Bacteria and mites

Cultures being studied and which are not preserved under oil or by lyophilization or liquid nitrogen storage are subject to contamination by mites and bacteria. Mites present particular problems as they are liable to move from one culture to another carrying spores of various fungi with them. Acaricides such as paradichlorobenzene which are used to keep insects away from herbarium specimens (see p. 31) may be kept with cultures to act as deterrents. Crystals of paradichlorobenzene added to infected cultures may kill mites but may also cause the fungus to deteriorate. For a review of acaricidal treatments see Smith (1967).

The most satisfactory way to keep mites out of tubes and bottles is by sealing them with cigarette papers which allow air to pass through them but have pores too small to allow mites through. About 25 ml of an adhesive, made up of 20 g gelatine in 100 ml distilled water to which 2 g copper sulphate have been added, are poured into a Petri dish and allowed to solidify. The cotton wool plug in the tube is flamed, pushed down, flamed again and while the rim of the tube is hot it is pushed into the gelatine mixture, removed, and immediately pressed on to the centre of half a cigarette paper (sterilized with propylene oxide). When the gelatine mixture has set the surplus paper can be flamed off to leave an unobtrusive seal.

Where bacterial contamination has occurred it is necessary to subculture the fungus on to a medium containing an antibacterial antibiotic (e.g. aureomycin, chloramphenicol). Details of the composition of some antibacterial media which permit the growth of fungi are included in Booth (1971c).

III. TAXONOMIC RANKS

A HIERARCHICAL system of classification is used in plant nomenclature based on the affinity of the different units. All available ranks (Table 2) need not be used

Table 2. *The principal taxonomic ranks*

Ranks	Examples		
Kingdom	Fungi	Fungi	Fungi
Division	Eumycota	Eumycota	Eumycota
Subdivision	Basidiomycotina	Ascomycotina	Ascomycotina
Class	Hymenomycetes	Pyrenomycetes	Discomycetes
Subclass	Homobasidiomycetidae	—	—
Order	Agaricales	Sphaeriales	Lecanorales
Suborder	—	—	—
FAMILY	Agaricaceae	Xylariaceae	Cladoniaceae
Subfamily	—	Hypoxyloideae	—
Tribe	Agariceae	—	—
Subtribe	—	—	—
GENUS	*Agaricus*	*Hypoxylon*	*Cladonia*
Subgenus	—	—	*Cladonia*
Section	*Agaricus*	*Papillata*	*Cladonia*
Subsection	—	*Primo-cinerea*	*Cladonia*
Series	—	—	*Cladoniae*
Subseries	—	—	—
SPECIES	*campestris*	*serpens*	*furcata*
Subspecies	—	—	*furcata*
Variety	—	*macrospora*	*palamaea*
Subvariety	—	—	—
Form	—	—	*rigidula*
Subform	—	—	—

in the classification of a particular individual and in most cases they are not; indeed I know of no instance in mycology where all ranks have been employed. The order in which ranks are used is fixed by the International Code of Botanical Nomenclature (Art. 4; see p. 133). In addition to the ranks formally recognized in the Code a few special ones outside the formal system are also used occasionally in mycology and lichenology. The most important of these are the 'special form', ('forma specialis'), 'modificatio', 'teratological form', 'chemotype', 'morphotype' and 'strain' categories.

A clear understanding of taxonomic ranks and their application is a necessary basis for all serious taxonomic study. The ranks summarized in Table 2 are used in all plant groups and not only in the fungi. It is consequently desirable that concepts of particular ranks should be comparable in different groups. Mycologists should be particularly conscious of the usage of the categories in vascular plants and endeavour to model their concepts on these. Students embarking on taxonomic revisions in mycology will find the works of Stebbins (1950), Davis and Heywood (1963) and Leenhouts (1968) invaluable introductions to the practice of taxonomy in vascular plants.

Notes on the usage of different ranks are presented below. Nomenclatural aspects, unless peculiar to particular ranks, are generally omitted here and discussed separately later (pp. 124–178).

Ranks above family

The ranks above that of family which are most frequently used in mycology are the subdivision (ending '-mycotina'), class (ending '-mycetes'), subclass (ending '-mycetidae') and order (ending '-ales'). These major divisions are ones of convenience and are based mainly on the way in which the spores are produced and the arrangement of the spore-bearing organs. In the subdivision Basidiomycotina, for example, the sexual spores are borne on basidia, whilst in the subdivision Ascomycotina they are produced within asci. The order Eurotiales is distinguished from the order Sphaeriales in that the unitunicate asci are arranged in cleistothecia and perithecia, respectively.

Ranks above that of family are not controlled by the International Code of Botanical Nomenclature (Art. 16; p. 142) and it is therefore not necessary to cite an author's name when using them. Such names are generally fixed by convention and consequently it is not uncommon for different authors to use quite different names for the same groups. Most names at these ranks are, however, fairly well established and the determination of a fungus represented by adequate material to the rank of order rarely presents any problems. Some authors are now attempting to apply the principles of typification, priority and valid publication to ranks above that of family but this practice has not yet been widely followed. Authors wishing to describe a new order, however, would do well to do so in Latin and indicate a type family as this fixes the application of the name and makes it intelligible to foreign workers who might not be familiar with the language used in the rest of the paper.

Families, subfamilies and tribes

The family is the unit of most concern to the majority of mycologists between those of genus and class. Each family consists of one or more genera which are considered to be closely related (monophyletic). Family names are formed from generic names and given the ending '-aceae' to agree with the feminine noun 'Planta', the Kingdom within which the fungi have until recently been generally placed (see Ainsworth, 1971; Ainsworth et al., 1973). A few authors regard the fungi as belonging to a Kingdom named the Protoctista (=Protista) (see Copeland, 1956) and if this view is followed family names must be given the ending '-acea' to agree with the neuter noun 'Protoctista'. In the fungi families are mainly separated according to the way in which the spore bearing organs are arranged and the tissue containing them (i.e. the sporocarp, if one is present at all). Family names are controlled by the Code and the pertinent Articles are included in the preliminary catalogue of family names proposed for the fungi and lichens compiled by Cooke and Hawksworth (1970). Changes in family names, like all other changes in nomenclature, should not be made unless they are impossible to avoid. If a well established name in this rank is found to be invalid or illegitimate or in need of rejection on the grounds of priority steps should be taken to see if it can be conserved before it is changed (see pp. 140–141). Relatively little attention has been given to names in this rank in fungi and lichens (with the exception of a few groups such as the Homobasidiomycetes) and it seems probable that many will eventually require conservation. Most well established family names used in the vascular plants were conserved to prevent any such unfortunate changes.

The ranks of subfamily, tribe and subtribe are rarely employed except in the larger families of fungi (e.g. Agaricaceae Fr., Xylariaceae Tul.). These contain groups of genera which appear to be more closely related to each other than to other similar groups within the same family but which are not separated by characters which would justify placing them in different families.

The authors of family names should be cited in the same way as the authors of generic and infrageneric names although many mycologists frequently omit to do this. A useful convention by which the suprageneric classification of a genus can be given concisely is to separate the suprageneric names by hyphens, in descending order of magnitude. *Hypoxylon* Bull. ex Fr., for example, could be referred to in parenthesis as '(Ascomycotina-Pyrenomycetes-Sphaeriales-Xylariaceae-Hypoxyloideae)' if introduced in a context where its systematic position needs to be made clear. This convention is also of value in diagnoses of newly described genera.

It is usual to place all suprageneric ranks in Roman (Agaricaceae Fr.) rather than italic (*Agaricaceae* Fr.) type.

Genera

Fungi are named according to the binomial system whose widespread adoption in botanical nomenclature dates from Linnaeus (1753; see Stearn, 1959). The first name by which any fungus is known is the genus. This name was formerly usually derived from a Greek word but may now come from any source (see p. 144), and is latinized, always spelt with a capital letter, and put in italic type (e.g. *Alectoria* Ach., from the Greek word 'ἄλέκτρος' meaning 'unwedded', from the rarity of the production of ascocarps in this genus).

The function of a genus is to group species into what are clearly homogeneous units based on a few carefully selected characters. What constitutes a particular genus is always a matter of debate and generic concepts vary markedly between different families. Ideally genera should be separated by several distinct unrelated characters. If the differences are relatively small or inconstant, infrageneric ranks such as 'subgenus' or 'section' might be more suitable. Where intermediate species are found separating relatively small genera it is often better to unite them rather than describe a new genus to accommodate the intermediates. Where 'intermediate' species are discovered separating large genera each containing hundreds of species, however, it is most convenient to keep the established genera distinct and place the 'intermediate(s)' in one of them. Remember that in many cases species thought to have been intermediate have later been found clearly to belong to one genus and not the other on the basis of characters which might not at first be readily apparent (e.g. chemical, onto-genetic, ultrastructural). In deciding whether or not to recognize a new genus care should be taken to ensure that the characters which distinguish it from other genera already accepted in that rank in the same family are comparable to those separating the accepted genera from each other.

Changes in generic names, particularly well established ones, are a nuisance as they often necessitate the re-arrangement of herbaria, additional cross-references in catalogues, the citing of the older name in parentheses, may lead to confusion in plant pathology, and provide additional names to be learnt. Such changes (e.g. raising sections to the rank of genus and *vice versa*) should only be made if really unavoidable and there are very strong grounds to sub-stantiate them (e.g. the discovery of previously overlooked characters). If in the course of a taxonomic revision an earlier validly published name at the rank of genus is found for a well established genus, the mycologist should not rush to make all the necessary new combinations into it without first attempting to have the name conserved under the Code (see pp. 130–140). A mycologist's taxonomic judgement will be more respected if he attempts to conserve well established generic names rather than introduce large numbers of new combinations under unfamiliar ones.

In selecting a name for a new genus care must be taken to ensure that it is not a later homonym (i.e. spelt exactly like another name based on a different type). It is relatively easy to ascertain if such a name exists in the fungi or lichens through Ainsworth (1971) and the *Index of Fungi*. Searches must also be made, however, to ensure that there are no other groups of plants (e.g. vascular plants, bacteria, algae) with names identical with it. It is embarrassing for an author who has just published a genus to find that his name has been validly published for a different group and to have to coin a new name and introduce the necessary new combinations shortly afterwards! To forestall this the principal catalogues of generic names of all major plant groups (listed on p. 109) should be checked. New names should be as dissimilar as possible from those already used to minimise confusion that might arise due to typographical and spelling errors. Cases like *Tichothecium* Flot. and *Trichothecium* Link ex Fr., and *Desmazieria* Mont. and *Desmazeria* Dumort. should clearly be avoided.

Several ranks are available between genus and species (Table 2). Subgenera, sections, subsections, series and subseries have latinized names which are similar to those of genera but with different endings. They have little application except in particularly large genera. The word 'series' has been used in a different manner by some mycologists, for example, in *Penicillium* Link ex Fr. by Raper and Thom (1949). Thus the '*P. pallidum* series' is used to include *P. pallidum* G. Sm., *P. putterillii* Thom, *P. namyslowskii* Zaleski and *P. lavendulum* Raper & Fennell. The usage of 'series' in this way corresponds most closely to the concept of 'aggregate species' or 'group' (see p. 43) of vascular plant taxonomists and should consequently be avoided.

Species
The second part of the binomial by which an individual fungus is known is

the specific name (epithet). The species name should be Latin in origin unless derived from a man's name, geographical location, or the generic name of a host when it is latinized. The ending of the specific epithet must agree with the gender of the generic name, e.g. *Alectoria jubata* (L.) Ach. (feminine) and *Bryopogon jubatus* (L.) Link (masculine) based on *Lichen jubatus* L. (masculine). The specific epithet should always be spelt with a small and not a capital letter. When an epithet is derived from a person's name some mycologists have used a capital letter (e.g. *Chaetomium ellisianum* Sacc. & Syd. or *Chaetomium Ellisianum* Sacc. & Syd.) but it is now recommended that a small letter be used in these cases as well. Species names should be placed in italic type just as generic names are unless the text is in italic type when they should be placed in normal type (i.e. 'The taxonomy of *Candida vulgaris* Berkh. is discussed', or '*The taxonomy of* Candida vulgaris *Berkh. is discussed*). The authority of the specific name (see pp. 159–163) should be cited at least the first time it is mentioned in any publication.

The concept of 'species' is the most fundamental a taxonomist must grasp and yet it is the most difficult to define and explain. Most experienced taxonomists in mycology would agree with the concept as used by each other but find it difficult to agree on a simple definition. The most generally accepted definition in vascular plants is probably that of Du Rietz (1930): 'The smallest natural populations permanently separated from each other by a discontinuity in the series of biotypes . . .'.* In practice the species should include individuals (i.e. samples from populations) separated from all other groups of individuals by marked discontinuities, preferably in several unrelated characters. In many groups of fungi relatively few characters are available and species are commonly defined on the basis of single well marked discontinuities. The interpretation of what constitutes such a well marked discontinuity must rest with the individual taxonomist and varies in different groups. The species within a genus should, however, be separated from one another by characters of as far as possible the same order. If a mycologist discovers individuals which appear distinct in some character from those described he should examine the described species already in the genus and see if the characters separating them from each other are equivalent to those distinguishing the individuals he has found from those already described; if they seem less important then a rank such as 'variety' below that of species might prove to be more appropriate.

In the choice of a name for a new species care must be taken to check that the name has not previously been validly published in that genus in the rank of species. Appropriate catalogues of names (see p. 108) should therefore be carefully checked. It is also advisable to endeavour to avoid the use of names in species rank applied in allied genera (which might later require transferring to the genus under consideration) and in infraspecific ranks within the same genus. In selecting a new species name it is recommended that one which is not likely to be misspelt (e.g. *Ascobolus crec'hqueraultii* Crouan should have been avoided;

---

* The term 'biotype' is defined by Jeffrey (1973) as 'A given genotype . . . expressed in one or more individuals'. 'Genotype' refers to the hereditary (i.e. genetic) constitution of an individual in contradistinction to 'phenotype', the outward appearance of an individual (i.e. the response of the genotype to environmental factors); see also Davis and Heywood (1963).

see also p. 148) and one which gives some idea either of its distinguishing characters (e.g. *Gelasinospora tetrasperma* Dowd. which has four-spored asci) or its
geographical distribution (e.g. *Alectoria himalayana* Mot. described from the
Himalayas) should be chosen.

The 'aggregate species' concept (e.g. *Trichoderma harzianum* Rifai agg., or
'aggr.') has been used in mycology, and is applied either where a taxon is
recognized as probably comprising a number of distinct species (or 'micro-
species' of angiosperm taxonomists) which require further work (i.e. as equivalent to '*sens. lat.*' or in the sense of a supraspecific rank in 'difficult' groups of
species (see Heywood, 1963; Davis and Heywood, 1963). The 'species aggregate'
concept is to be avoided as it is merely a method of avoiding the issue of what
rank to apply to particular populations or making more critical observations.

'Stirps' (Stirpes), first used by E. Fries in 1862, will sometimes be found
appended before a specific name. This word was once used to denote natural
groups of closely allied barely distinguishable species within genera and its
concept is essentially that of the 'aggregate species'. The word 'group' (e.g.
Rhizocarpon geographicum (L.) DC. group) is sometimes used in the same way
as that of the 'aggregate species' or 'stirps'.

Subspecies

This rank has been used in very many different ways in the past but has now
come to have a much more precise meaning. It has been little used in recent
years in the non-lichenized fungi but is employed in the lichens. In vascular
plants this rank has been defined as '. . . a population of several biotypes forming
a more or less regional facies of a species . . . The various subspecies of a species
are continuously intergrading into each other, their delimitation thus being
much more arbitrary than that of the species' (Du Rietz, 1930). The 'subspecies'
rank is used where two or more populations separated either geographically,
ecologically or both throughout most of their range and distinguished by
characters which might be used as criteria at the rank of species have intermediates where their distributions overlap (i.e. where they are sympatric) so
that it is not possible to place some individuals satisfactorily in one subspecies
or another. For animals the '75% rule' (see Mayr *et al.*, 1953; Edwards, 1954)
has been applied (i.e. 75% of the individuals must be distinct from other individuals of different subspecies) but in many vascular plant and lichen subspecies
a figure of 85–90% would give a better indication of its use in practice.

Subspecies are indicated by the use of either 'subsp.' or 'ssp.' between the
specific and subspecific names (e.g. *Cladonia arbuscula* subsp. *beringiana* Ahti or
C. arbuscula ssp. *beringiana* Ahti). It should be noted that the practice of not
using any indication of rank between specific and subspecific names which is
adopted in zoology (e.g. *Erithacus rubecula tartaricus*, *Troglodytes troglodytes
zetlandicus*) is not now followed in the plant kingdom. Bacteriologists treat the
categories of subspecies and variety as of the same rank (see Jeffrey, 1973).

Variety

The variety ('*varietas*'; 'var.') is extensively employed in the lichens and to a

lesser extent in the non-lichenized fungi and is defined by Du Rietz (1930) as
'. . . a population of one or several biotypes forming a more or less distinct facies
of a species'. Varieties are based on characters which are not regarded as meriting
separation at the rank of species or subspecies and which are distinct from all
other varieties within the species (i.e. intermediates should not occur). This rank
is commonly used in mycology for populations differing consistently in some
character such as spore size (e.g. *Thielavia terricola* (Gilman & Abbott) Emmons
var. *terricola* with ascospores mainly 12–17 × 7–10 μm, and var. *minor* (Rayss &
Borut) C. Booth with ascospores 9–12 × 5–6 μm). It is also commonly applied to
cases where there are either geographical differences of a more local nature than
those considered under the rank of subspecies, or ecological differences (e.g.
Haematomma ochroleucum (Neck.) Laund. var. *ochroleucum* with usnic acid is
mainly confined to rocks and walls in Britain, and var. *porphyrium* (Pers.)
Laund. without usnic acid occurs frequently on trees as well as on rocks and
walls; *Pseudevernia furfuracea* (L.) Zopf var. *furfuracea* with physodic acid is
commonest in southern Europe while the var. *ceratea* (Ach.) D. Hawksw. with
olivetoric acid is commonest in northern Europe).

The rank of 'chemovariety' ('chemovar.') was proposed originally by Tétényi
in 1958 for use in flowering plants for races distinguished chemically which
appear to merit separation at the rank of variety (see Tétényi, 1970). It has
occasionally been used in the lichens but should be avoided as it implies that
chemically characterized taxa are somehow inherently 'different' from those
defined on other grounds (see Lanjouw, 1958).

The term 'cultivar' is extensively used in agriculture and horticulture for
genetically distinct facies of species which have arisen in cultivation and the
nomenclature of cultivars has its own Code (Gilmour *et al.*, 1969). This rank
has been used in some microfungi (see p. 45) but is only correctly employed in,
for example, the case of cultivated mushrooms.

Form

The form ('*forma*'; 'f.'; 'fo.') is probably the most misused rank in taxonomy.
It has often been applied to any minor morphological variant regardless of the
source of the variation but has now come to be accepted as '. . . a population of
one or several biotypes occurring sporadically in a species-population (not
forming distinct regional or local facies of it) and differing from the other biotypes of this species-population in one or several distinct characters' (Du Rietz,
1930), and is often applied to differences which might well be due to a single gene.
In most cases this rank is probably best avoided except where the source of the
variation seems clear. In mycology it is also liable to confusion with 'special
forms' discussed below.

Special form

The special form ('*forma specialis*'; pl. '*formae speciales*') is a special rank
used principally in the fungi but occasionally in parasitic vascular plants (e.g.
Arceuthobium Bieb.; see e.g. Hawksworth and Wiens, 1972) and bacteria, and is
abbreviated to 'f. sp.' (plural 'ff. sp.' or 'f. spp.'). The abbreviation 'f.' which

has been adopted for *formae speciales* by some authors is not acceptable as this leads to confusion with the rank of 'forma' (see above). These special forms are characterized by their physiological reaction to particular host plants and their nomenclature, like that of the ranks discussed in the following paragraphs, is not considered by the Code (Art. 4; p. 134). This adaptation is essentially physiological but may also be accompanied by slight morphological differences. The name given to special forms is usually based on the generic name of the host, e.g. *Puccinia graminis* Pers. f. sp. *tritici* Eriks. & Henn. on wheat (*Triticum*) and *P. graminis* f. sp. *avenae* Eriks. & Henn. on oats (*Avena*). Even though these names are not controlled by the Code many mycologists now adopt similar criteria when publishing new special forms and give them author citations. 'Cultivar' names have sometimes been used within single *formae speciales* (e.g. *Fusarium roseum* Link f. sp. *ceralis* 'Avenaceum' and *F. roseum* f. sp. *cerealis* 'Equiseti'; see Booth, 1971*d*) but this interpretation constitutes an incorrect usage of the *formae speciales* concept (see Ainsworth, 1962; Hudson, 1970). If the races of a fungus on different cultivars are physiologically distinct then each should be given separate *formae speciales* names. A useful review of host specialization as a taxonomic criterion is given by Johnson (1968).

Morphotype, chemotype and ecotype

The two terms morphotype and chemotype have been given precise meanings by Santesson (1968). They are applied to populations of undetermined taxonomic rank or of no taxonomic value and are not given latinized names. Chemotypes are chemically characterized parts of morphologically indistinguishable populations whilst morphotypes are parts of morphologically characterized populations. The morphotype and chemotype concepts have wide applications in taxa which show morphological variation in response to various (often undetermined) factors and produce chemicals sporadically, respectively. In the case of chemical variation, particularly the simple presence or absence of an unreplaced compound not correlated with geographical or ecological differences, one can refer to, for example, the 'usnic acid present chemotype' of a species or its 'usnic acid absent chemotype' (see e.g. Hawksworth, 1970*b*). Many of the described 'forms' in lichens are probably more correctly regarded as morphotypes without any taxonomic significance. The term 'morph' is sometimes applied to distinct morphological types by North American authors. Huxley (1940) introduced the term 'paramorph' for any deviation from the mean of the group not meriting formal taxonomic recognition.

Vascular plant taxonomists make extensive use of the term 'ecotype' for *genetically* adapted races of species restricted to very precise ecological conditions. The adaptation may be morphological or physiological. The usage of this term is discussed in some detail by Heywood (1959) and Davis and Heywood (1963). Although the 'ecotype' concept has not yet been extensively employed in mycology it is likely to be increasingly used in the lichens as the adaptation of populations to particular habitats becomes better understood (see Hawksworth, 1973*b*). In many cases, in groups such as the lichens, great care will always be needed so as not to confuse 'ecads' (morphotypes adapted to an ecological niche through phenotypic plasticity) with 'ecotypes'. Ecotypes are essentially experimental units and may have names in accepted ranks (frequently as 'subspecies' or 'varieties'). Special forms might also be interpreted as ecotypes where their specificity has been proved experimentally.

Strain and Race

Strains of some fungi which have been grown and studied in culture for many years have been found to produce individuals with different genotypes. These mutant populations have been given numbers or codes in some species to denote these genetic differences (e.g. *Aspergillus nidulans* (Edam) Wint. C. bi 1; W3; S9). There is no agreed notation for denoting mutants of this type but the systems established in the genus or species within which one is working should be adhered to.

'Chemical strain' and 'chemical race' have been used in the lichens for chemically differentiated races of morphologically identical species. Each chemical strain can be numbered (e.g. *Cladonia acuminata* (Ach.) Norrl. Chem. str. II). As different authors occasionally use the same numbers for different races of a single species this type of notation can lead to confusion and should consequently be avoided and either one of the formal taxonomic ranks or 'chemotype' (see p. 45) used to indicate this type of variation.

The terms 'physiological strain' and 'physiologic races' are also encountered from time to time and refer to races characterized by particular physiological or pathological traits which may be genetically determined. 'Physiologic(al) race' has been extensively used in plant pathology but applied in various senses. The Federation of British Plant Pathologists (1973) define it as a taxon of parasites characterized by specialization to different cultivars of one host species. Robinson (1969) recommends the term 'pathotype' and 'physiotype' for pathologically and physiologically differentiated races of fungi, respectively, but the '-deme' terminology may be preferable (see Davis and Heywood, 1963; Federation of British Plant Pathologists, 1973); i.e. 'pathodeme', 'physiodeme'.

Different workers have employed different schemes for designating races within species and, as in the case of the terminology of mutant strains, the accepted system in various groups should be adhered to. For further information on the designation of physiologic races see Tarr (1972) and Gilmour (1973).

Modification

The modification (*'modificatio'*), abbreviated as 'mod.', has been used in the lichens and is usually employed for environmentally induced morphological variations with no genetic basis which do not require formal taxonomic recognition (see e.g. Weber, 1968). These have sometimes been given latinized names (most often as combinations from other ranks such as 'variety' or 'form') but this practice is not now usually followed and such names in any case fall outside the scope of the Code. Most described 'modifications' would now be regarded as 'morphotypes' (see p. 45).

Teratological form

Teratological forms ('terat.', 'ter.', 'T.', or 't.') are abnormalities which occur from time to time within a species. In some cases these abnormalities may have some genetic basis but in most instances they probably do not and consequently do not merit any recognition in an accepted rank. Grummann (1941) proposed a series of terms for the major types of such abnormalities which occur in lichens, e.g. T. *tortus* for spirally twisted fruticose pendent lichen thalli, but these have not been widely used by most taxonomists. Although new teratological forms are described from time to time (e.g. *Physcia biziana* T. *excrescens* Gallé described in 1959) they have little place in formal taxonomy although they may be useful in discussions of abnormalities which recur in, for example, several species of the same genus.

IV. NAMING, DESCRIBING
AND PUBLISHING

NAMING

IF A specimen has been collected and examined as discussed above (pp. 13–25) sufficient data will be available to start the task of ascertaining its correct name. The student who has either attended a course of lectures in mycology or read one of the recommended introductory texts (see p. 11) will have no difficulty in finding the major group to which his collection is assigned. The subsequent procedure is determined by both its group and country of origin. Literature pertinent to particular groups and regions may be traced through Ainsworth (1971) and the *Bibliography of Systematic Mycology*. Keys to most accepted fungal genera are included in the recent work of Ainsworth *et al.* (1973). Lists of publications of value for the determination of British fungi and lichens have been compiled by Holden (1969) and Hawksworth (1970a), respectively. The taxonomist working outside an institution with a comprehensive mycological library will first of all need to build up a collection of the major papers relevant to his group(s) and (or) region(s).

With the aid of such books and papers as are available an attempt to name the collection is made. At first this may prove a tedious and time consuming chore, but as familiarity with keys, the literature, and the characters of different taxa increases, the task becomes less arduous. When a name seems probable from the published description, comparisons with correctly named herbarium material or cultures are necessary. Bisby (1953) noted that young taxonomists tend to fall into one of four main categories; those who (1) accept any name without critical thought or study; (2) 'force' a specimen to fit a description; (3) generally find their material does not fit any description; or (4) proceed cautiously making tentative identifications and noting discrepancies from published descriptions. Students of the last type are the most likely to make valuable contributions to systematic mycology.

Unfortunately the majority of newcomers to mycology do not have access to either adequate libraries or comprehensive herbaria. Such students may prefer to proceed by drawing up descriptions and making sketches of their collections, and grouping them into what appear to be homogeneous 'species'. Each may then be allotted a temporary label such as a letter (e.g. '*Hypoxylon* A'). To find out if the material has a name a mycological herbarium and library must be visited or the material submitted to a specialist. The former course is preferable as only in this way will the mycologist become familiar with the literature and the characteristics of other taxa. A name written on a packet by a specialist is not as likely to lead to an understanding of the group.

Adequate published keys do not exist to many fungi and it will often be necessary to construct keys to particular families and genera. When the generic name has been ascertained with a reasonable degree of certainty lists are prepared of the species described in that genus (see p. 63). The descriptions of these can be looked up and notes prepared so that a key can be made (see pp. 71–75). Such herbarium material as is available should also be examined at this stage. As noted by Mason (1940) '. . . the surest basis of the art of diagnosis . . . is the matching of good specimens . . . against good specimens that have been correctly named.' Some fungal genera are so large that it would be too time-consuming to do this; in these cases taxa from the same host and (or) geographical area should be considered first. For microfungi host indices are a particularly valuable aid and these are usually provided in major taxonomic works and catalogues on these groups. The separate host index maintained at the CMI for the material in the herbarium (see p. 31) receives extensive use.

Specimens which are adequate but do not seem to fit any of the published descriptions or to be conspecific with any of the herbarium material should not be discarded, as many beginners tend to do, but retained for further study. 'Intermediate' collections may be significant and provide evidence for the union of taxa hitherto treated separately. Seemingly distinct specimens which do not agree with any of the published descriptions may either be undescribed altogether or described, perhaps incorrectly, in another genus.

The literature on fungi is so large and complex that only a specialist is likely to have collated all the available data on the taxonomy of a particular group. For this reason when the student has proceeded as far as he is able the material should be submitted to a specialist for his opinion. Any specimens sent to a specialist must be adequate in size and fully documented (see p. 29). Where possible a duplicate should be retained by the sender and the larger part sent to the specialist for his herbarium; care should be taken to ensure that the same organism is present on both parts of split collections. When dealing with pathogenic species care must also be taken not to send living cultures or specimens of species to countries where they may not be indigenous; if in doubt only dead material (i.e. dried, fumigated specimens or cultures) should be sent.

Sometimes material has been submitted to two or more experts at the same time. This practice is wasteful of specialists' time and is not to be recommended. If he is uncertain of an identification the expert will make this clear to his correspondent and if a second opinion is necessary suggest to whom he should submit his collection. Specialists are usually in contact with other workers concerned with their group(s). Taxonomy is continually in a state of flux and because of this different mycologists may have their own views on the delimitation of certain taxa. If a determination seems dubious for any reason the mycologist should be approached and asked to provide further information. It may be that some feature was glanced over too rapidly by an overworked specialist! Many mycologists appreciate receiving notes (e.g. spore sizes, chemical tests), slides and sketches accompanying collections to reduce their burden; these also demonstrate that the sender is in earnest. If a specialist feels that he is dealing **with a keen and promising student he will naturally be more inclined to assist** him by providing notes as to why, for example, the student's provisional identifications were incorrect! Too many collections should not be sent at one time; less than ten usually ensures fairly prompt replies. The student should not expect rapid replies as specialists often have many other commitments. Experience of dealing with particular specialists will show how best they like their collections annotated, as many will indicate what information they require. The British Lichen Society has a panel of 'referees' to assist members in naming their collections, and the American Mycological Society and the British Mycological Society have comparable lists of specialists willing to help in the determination of other fungi (see also Shelter and Read, 1973).

Although some institutions have specialists dealing with all major groups of fungi most do not, and the student will find that he has to send collections of different groups to a number of experts. Bisby's (1953) hope for a 'clearing house' for the forwarding of specimens to specialists remains a very desirable but still unattained goal.

The beginner should proceed cautiously and learn groups thoroughly before attempting ambitious projects so that his generic and species concepts are clear. There are so many undescribed fungi that before long something which appears to be new to science will be discovered. He must be wary of rushing into print, however, and remember that probably less than one in two 'new' fungi are in fact previously undescribed, and that of these, probably one third to a half are described in an incorrect genus.

The following sections discuss in further detail the description of fungi and the preparation of systematic papers for publication. The selection and application of names in the fungi is discussed on pp. 124–178.

DESCRIBING

EVENTUALLY the mycologist will have to draw up a description of a fungus. A 'description' includes all characters while a 'diagnosis' only those which distinguish it from others. In monographic and revisionary studies it is not always necessary to cite characters common to all species of a group if these are presented in a description of the next higher taxon to which they belong (e.g. the genus or section).

The most satisfactory way to start drawing up a description is to list all characters down one side of a table and then proceed to score the character states for each specimen or taxon in separate columns. This tabular method of recording descriptive data not only ensures that all characters are investigated for each specimen or taxon but is very helpful in the construction of analytical keys (cf. p. 73). As many specimens as possible should be studied. Ideally a description of a species in a monograph should cover the entire world population; past, present and future (Mason, 1940). This is, of course, an unattainable goal, but authors might do well to consider its implications. Mycologists should not, however, continue to accumulate descriptive data indefinitely; when they find that the characters of additional specimens studied fall within the range of variation already recorded it is clear that the data are then reasonably comprehensive.

When sufficient data are assembled the writing of the description may be started. Each organ is taken in turn in a logical sequence and its full range of variation described in detail. The first characters to be described are usually the gross ones and microscopic features are dealt with in the following sentences and phrases. The life-form of a lichen thallus, the pileus of an agaric, and the appearance of a culture are examples of suitable starting points. Each character is followed by a list of appropriate adjectives and measurements separated by commas and, where necessary, semi-colons. It is a good policy to study recent descriptions of taxa in the same group and, where these seem comprehensive, to draw up descriptions which are comparable, as this will facilitate comparison by other workers. Unfortunately students often omit to mention characters other specialists consider vital and would have liked information on.

Where several descriptions of allied taxa are included in a single publication they should always be comparable, i.e. take characters in the same sequence and describe them in parallel ways. If a character is absent in one taxon but present in others of the same genus it should not simply be omitted but clearly stated to be absent (e.g. '*Chlamydospores* absent', '*Soralia* absent').

It is usual to subdivide descriptions into several short paragraphs each dealing with related characters (e.g. gross form; soralia; perithecia, asci and ascospores; conidiophores, conidiogenous cells and conidia). To make cross-checking quicker many authors now place the name of each principal character in italic type and the description in Roman type; i.e. in the form '*Conidia* ellipsoid, hyaline, 1 septate, $8-12 \times 6-9$ μm'. Some authors also emphasize those characters which are diagnostic by printing them in either italic or spaced Roman type; e.g. 'Ascospores hyaline, *3-4 septate*, $10-15 \times 5-7$ μm' or 'Ascospores hyaline, 3-4 s e p t a t e, $10-15 \times 5-7$ μm'.

The latter part of a description often includes information on chemical components, habitats (e.g. substrate, host range), pathogenicity and geographical distribution. Where these data are not diagnostic it is clearly preferable to separate them from the main description.

In presenting any description it must always be made clear just what material has been examined in the compilation of the description. Where rather few specimens have been available for study full details of them can be presented but where many have been examined this will not be practical and a list of representative specimens studied can be included instead. For information on the citation of specimens see pp. 68-70. Descriptions included in regional studies such as floras should always be based only on material from that region although in them it may be helpful to indicate variations noted in adjacent regions as long as it is made clear that they do not occur in the area being investigated. In a study of a lichen sterile in Britain in a paper on the British population, for example, a phrase such as '*Ascocarps* (unknown in Britain; seen in North American specimens)', may be used to introduce a description of ascocarps based on foreign collections.

In some instances authors find either that their observations are not in accord with ones already published, or that particular character states are absent in their specimens. The tendency to copy information from previously published descriptions allegedly of the same taxon must be avoided at all costs. Such observations should, however, be referred to, and this can be done in descriptions themselves by including these data in parentheses, e.g. 'conidia 15–22 × 4–6 μm (20–24 × 6–8 μm *fide* Höhnel, 1909).' Where considerable discrepancies are discovered between descriptions of taxa purporting to be the same the problem should be investigated in detail by studying the collections and (or) drawings of the author concerned where this is possible.

It is essential that descriptions of fungi and lichens are adequately illustrated by line drawings and(or) photographs which clearly show their diagnostic characters. The illustration of mycological specimens is discussed in the following section (pp. 57–62).

Great care is required at all stages in the drawing up of any description as any errors may mislead future workers. There are already too many examples of inaccurate and confused descriptions in the mycological literature. For an example of the confusion which can arise see Lentz and Hawksworth (1971). When a description has been completed the manuscript should be checked carefully both against the original table of data and representative specimens so as to ensure that it is entirely accurate. In the process of retyping manuscript descriptions particular attention must be paid to ensuring that the measurements are correctly typed.

Terms

Many groups of fungi have developed specialized terminologies to facilitate precision in their description. Where such terminologies are available and employed attention must be paid to their precise meaning and glossaries including definitions of them compared. Definitions of most of the commonly employed terms in mycology are included in Ainsworth (1971) and in the glossary of Snell and Dick (1971). Most terms related to conidium ontogeny are discussed in Kendrick (1971). Jackson (1928) and Swartz (1971) provide comprehensive glossaries of general botanical terms. In some cases the same term has been applied in different senses by different authors and these should be avoided unless their meaning in a particular context is made quite clear. Where there is any possibility of confusion it is helpful, particularly in larger floristic and monographic studies, to define and illustrate them clearly either in an introductory chapter (cf. Ellis, 1971) or in a glossary (cf. Duncan, 1970; Hale, 1969).

Where several alternative terms are available for the description of a particular character state, that most widely used is to be preferred. Authors should endeavour to make their descriptions as readily comprehensible as is compatible with precision to mycologists who are not specialists in the group without their having continually to refer to either a glossary or Ainsworth (1971). Mycologists are unfortunately as prone as more general readers to 'gloss over' words or phrases they do not fully understand.

Shapes

The terminology of simple symmetrical shapes has been investigated by the Systematics Association Committee for Descriptive Terminology (1960, 1962) who proposed a standardized list of terms with English, French and Latin equivalents. Solid shapes are of particular importance to mycologists and Ainsworth (1971) provides illustrations of many of these based on spheres and ellipsoids.

Measurements

Measurements constitute an essential part of descriptions of all groups of fungi and lichens and are now usually given in metric (S.I.) units, e.g. m, cm, mm, μm (see Anon., 1972; British Standards Institution, 1969; Royal Society, 1971). The symbol 'μ' ('micron') is still used in many journals instead of 'μm' but it is not now advocated by the S.I. System. Various other units have been employed by mycologists in the past of which the British inch (0·0001 ins = 2·54 μm) and Paris inch (0·0001 Paris ins = 2·79 μm)* are the most frequently encountered. The temptation to replace figures by phrases such as 'size of a pea' is to be avoided as are categories such as 'large', 'small' and 'very small' unless they are defined with reference to particular measurements in the same publication (cf. Orton, *in* Henderson *et al.*, 1969).

Too much 'precision' can give a misleading impression of accuracy. '50·27 cm' should be replaced by '50 cm' and '1·271 μm' by '1·5 μm'. Bisby (1953) recommends that all measurements made with an eyepiece micrometer should be given to the nearest 0·5 μm as the 'personal factor' in an observation will often exceed this value as will the variability of the object being measured. The 'personal factor' can assume considerable importance when different authors give significantly different measurements for the same material. Harter (1941) has emphasized the importance of the 'personal factor' in studies on the genus *Fusarium*.

When describing taxa as many measurements as possible are made. It is not sufficient merely to take the 5–10 recommended for routine examination of spores (see p. 23) and 50–250 are usually necessary, depending on the group concerned and the number of specimens available, to provide an accurate figure for the spore range of a species. Measurements of immature structures should be omitted unless the author makes it clear that his data refer to immature states.

In mycological publications measurements are generally presented either in the form '(8–) 10–12 (–15) × (3–) 4–5 (–6) μm' or '8–15 × 3–6 μm'. The first set of figures in each case refers to the longest, and the second to the shortest axis of the structure. For spherical structures only a single set of measurements is required. Figures in parentheses in the first example are the extremes recorded while those not in parentheses indicate the size range of the majority of structures (usually about 90%). Some authors give only single figures, the arithmetic means of all measurements taken on each axis, but this is clearly not satisfactory as the reader is then unaware of the range of measurements involved; '9 × 4 μm' might indicate either '8·5–9·5 × 3·5–4·5 μm' or '6–12 × 2–6 μm'. An alternative

* See also Ainsworth (1971, p. 327).

which is occasionally employed is to use the arithmetic mean and indicate the limits in parentheses, e.g. '(6–)10 (–12) × (2–) 4 (–6) μm'. Authors should ensure that if they use a system other than either the '(8–)10–12(–15) × (3–) 4–5 (–6)' or '8–15 × 3–6' type it is clear what their measurements represent. A few authors now include statistical information in their citation of measurements (see e.g. Sheard, 1973).

Colour

Accurate description of colour is essential in mycology, particularly in the Agaricales where considerable emphasis is placed on this character in the delimitation and determination of species. Various colour charts are available to mycologists and Ainsworth (1971) provides a list of these. The specimens are matched by eye with coloured patches and the best possible fit obtained. Some of the colour names employed in colour charts may be unfamiliar to students not conversant with them and so it is advisable in cases where the colour is of particular taxonomic significance to give it both in general terms and by reference to the chart employed, e.g. 'yellowish-grey-green (pistachio green, Munsell 2·7G/5·6/3·2)', 'pale yellow (pale luteous, Munsell 2·4Y/8·5/7·0)'.

Some of the older colour charts (e.g. Ridgway, 1912) are not now satisfactory as copies in different institutions have assumed differing colours over the years. Colours are currently most satisfactorily standardized by reference to the *Munsell Book of Colour*. The charts of Rayner (1970) and Kornerup and Wanscher (1967) provide Munsell notations and are suitable for most mycological purposes. Dade (1949), Rayner (1970) and Stearn (1973) provide English and Latin equivalents of colour terms.

Colour is inherently a rather variable character and attention should therefore be paid to the range of variation encountered. Not only may the colour of fresh material vary but sometimes colours change markedly when material has been in the herbarium for some time. Some lichens which are greyish-green in the field, for example, assume a much brighter yellow colour in the herbarium (e.g. *Alectoria ochroleuca* (Hoffm.) Massal.). Consequently it is essential that it is clear whether any colours given refer to fresh or herbarium material and where changes are known to occur this should be stated (e.g. 'pale grey, becoming reddish in the herbarium'). The colour of bruised and cut tissues in some Agaricales varies from that of undamaged tissues and this difference also must be made clear (e.g. 'pileus fawn, becoming blue-green when bruised').

A single colour chart and notation should, of course, be employed consistently throughout any particular publication.

Cultures

Many fungi are now routinely examined in culture and so where the species is known to grow in culture details of the cultural characteristics form an important part of their description. The media employed may affect the form of a colony markedly. Some *Chaetomium* Kunze ex Fr. species, for example, produce abundant aerial vegetative mycelium on PDA but no clearly apparent vegetative mycelium on PCA. Growth under near-ultraviolet (UV) light may influence the production of sporocarps and the septation of conidia. Temperature has a profound effect on some fungi; a species may, for example, fail to produce colonies at 20°C but grow vigorously under otherwise identical conditions at 30°C. For these reasons the medium and precise conditions under which a species has been grown must always be stated. If these data are not presented the value of the subsequent information will be reduced considerably.

For the description of cultures subcultures on plates (Petri dishes) are the most satisfactory. Attention must be paid to (a) the texture, (b) shape in section (e.g. convex, tufted), (c) nature of spore masses in Hyphomycetes (e.g. slimy, dry), (d) the time taken to form sporocarps or characteristic resting structures (e.g. chlamydospores, microsclerotia), (e) colour from above, (f) colour from the reverse of the plate (i.e. including pigments diffusing into the medium), (g) any odours, (h) any tendency to form sectors, and (i) growth rate (e.g. mean time values taken to form a colony with a particular diameter based on several replicates). Particular care is needed in describing the colours of colonies as in some genera these are liable to change (e.g. darken) as the colonies age.

Where possible information on the cultural characteristics of the species on a variety of media under a range of environmental conditions should be provided. It is not uncommon for authors to devote 6–10 pages to accounts of the cultural and(or) physiological characteristics of a new taxon (e.g. Ware, 1933; Isaac and Davies, 1955).

Chemistry

Chemical characteristics play only a relatively minor role in the taxonomy of most groups of non-lichenized fungi. Watling (1971) provides an account of chemical tests applied in the Agaricales and precautions to be borne in mind when employing them. Iodine is widely used in some groups of the Ascomycotina.

In the lichens, however, no modern descriptions should overlook the chemical components. These are normally listed towards the end of the descriptions and comprise both the colour reactions with standard reagents (i.e. C, K, KC, PD*) and the names of the compounds when these have been determined. Positive and negative reactions should always be indicated for each of the standard reagents, even if all are negative. 'Plus' and 'minus' signs are used to indicate positive and negative reactions respectively, and 'plus' signs are succeeded by the names of the colours (e.g. K + yellow, K−). Where colours change this is indicated either in the form 'K + yellow → red', or 'K + yellow > red'. Colours have often been abbreviated to their initial letter (e.g. K + y, KC + r) but this is not recommended unless the colour abbreviations employed are defined separately, as they may lead to confusion. Different tissues of the lichen may give different reactions with the same reagents (e.g. epithecium, cortex, medulla, thecium) and so it must always be made clear to what tissue the tests refer (e.g. 'Cortex K + y, C−, KC−, PD + yellow; medulla, K−, C + red, KC + red, PD−'). The convention 'K$\frac{+}{-}$' has been used where the upper symbol refers to the cortex and the lower to the medulla but can lead to confusion; 'K ± yellow' might mean

* See p. 20 for details of composition.

either a variable reaction or 'Cortex K + yellow; medulla K −'. This convention is fortunately rarely used now.

The identities of the chemicals can either be placed in parentheses after the reaction they give rise to (e.g. 'C + red (olivetoric acid)') or listed after all the reactions. It is essential to provide details of the techniques used to determine the chemicals reported; crystal tests alone are rarely satisfactory and thin-layer chromatography (TLC) is currently used routinely (see pp. 24–25). Data on unidentified compounds should be presented in full.

When examining material closely adhering to bark for terpenes by TLC, bark-extract alone should be studied on the same TLC plates to forestall host terpenes being reported from the lichen.

As some taxa vary in their chemical components as many specimens as possible should be investigated. Some correlations with ecological, geographical or morphological characters may be found to occur. The value of chemical characters in taxonomy is further discussed below (pp. 98–100).

Latin

The validating diagnosis of any taxon between the ranks of family and forma (inclusive) published after 1 Jan. 1935 must be in Latin or refer to a previously validly published diagnosis (see Art. 36; p. 156). The Latin diagnosis may be a translation of the entire description or an abbreviated form including only the characters separating it from allied taxa. A description of this latter type may merely be a single line.

Stearn (1973) points out that Botanical Latin has evolved independently from Classical Latin over the last 250 years in response to requirements of precision and economy of words. The divergence is so marked that Botanical and Classical Latin are most appropriately regarded as separate languages. All mycological taxonomists need some knowledge of Botanical Latin for, even if they never have to describe a new taxon, there will be occasions when they have to refer to those published by other authors. A comprehensive account of Botanical Latin is provided by Stearn (1973) who includes examples taken from both lichenology and mycology. It is essential that students use this work extensively when drawing up Latin descriptions. Valuable mycological-Latin glossaries are provided by Cash (1965) and Clements and Shear (1931). Woods (1966) provides Greek and Latin equivalents of terms employed by plant taxonomists which may be useful in coining new generic and specific names; the work of Jaeger (1955) is also of some value in this connection. Latin-German equivalents of place names are compiled in Graesse et al. (1971). Special glossaries dealing with colours and shapes have been referred to above.

In constructing Latin descriptions attention must be paid to the precise meanings of words ensuring that they are employed in accordance with current mycological practice. Where there is any doubt about the meaning of a particular word a standard Latin dictionary (e.g. Lewis and Short, 1955) is necessary. Before embarking on the Latin translation, however, it is advisable to study carefully the original descriptions of recently described taxa in the same or closely allied genera.

When the Latin description has been drafted it should if possible be given together with the English version to a colleague more familiar with Botanical Latin (not necessarily a mycologist) to ensure that the precise meanings are conveyed and that the whole is grammatically correct.

If a monotypic genus is being described for the first time a combined generic-specific description is permitted under the Code (Art. 42; p. 157). Some authors, however, prefer always to provide separate descriptions to make it clear what characters they consider to merit recognition at the rank of genus.

Some authors provide an original description which is only of the holotype specimen although many have been studied. This practice has been recommended (cf. Leenhouts, 1968) but has not been generally adopted by mycologists. Where such a procedure is employed two Latin descriptions are advisable, '*descriptio typi*' dealing with the holotype specimen alone (see p. 127), and a '*descriptio specie*' dealing with the range of variation in the species as a whole. This method may prove useful if the holotype is lost as it may enable a more satisfactory lectotype or neotype to be chosen (see pp. 127–128), particularly in cases where the 'species' is later found to consist of two or more taxonomically distinct elements.

ILLUSTRATING

ILLUSTRATIONS (drawings, graphs, maps, plates and tables) form an essential part of taxonomic accounts as they are able to convey considerable amounts of data which may not be as readily apparent from the written text. A good line drawing, for example, can be more informative than a whole page of detailed description.

Illustrations are often expensive to reproduce and for this reason most journals are only prepared to publish those which are of a high standard and essential to the paper. For each taxon both macroscopic and microscopic characters should be illustrated, paying particular attention to the diagnostic features. Two main types of illustration are available to mycologists, line drawings and photographs, and both have their own merits. Photographs have the virtue of authenticity and are particularly valuable for macroscopic illustrations. They are less satisfactory, however, for microscopic structures which are not dispersed in a thin optical plane, and more expensive than line drawings to reproduce. Line drawings, in contrast, are necessarily less realistic but when constructed accurately often prove more informative than poor quality photographs (particularly in the Hyphomycetes). In practice many mycologists prefer to use a combination of line drawings and photographs.

Drawings and photographs require to be set up into 'plates' before they are sent to an editor, although some journals now require authors to submit two sets of photographs (one mounted and one unmounted). Before starting to arrange the items it is necessary to measure the page size of illustrations in the journal to which the article is to be submitted and, remembering to allow space for any legends that may be required, to make up the illustrations either at that size or at multiples of it suitable for reduction by, for example, $\frac{1}{2}$ or $\frac{1}{3}$. The individual components of a plate may be trimmed most satisfactorily with a guillotine and dry-mounted on to stiff white card (e.g. Bristol board). Photographs are best mounted slightly apart so that a space of about 1–2 mm will separate them at the size of final reproduction. Adequate margins should always be left around the originals of each illustration (4 cm is recommended) and authors should indicate in pencil in a margin their name and the number of the figure or plate. Figures and photographs are often lettered by a journal's printers and authors should check if they have to add letters or not for the journal they have in mind. Line drawings can easily be lettered in light pencil but this is not possible for photographs where a pencil sketch will be required. If the lettering has to be done by the author care must be taken to ensure that it will not be too large or too small when reduced to page size. Both self-adhesive letters (e.g. Letraset) and stencils (e.g. Uno Stencils) are adequate for this purpose; white grounds or white letters may be necessary for labelling dark sections of photographs. Electric typewriters produce most satisfactorily any longer labels that might be required.

Scales may be drawn directly on to line drawings but often look untidy if drawn directly on to photographs. Scales are always given in a round number (e.g. 10 μm not 9·5 μm) and their length corrected accordingly. In the case of photographs, magnifications are most appropriately incorporated into the legends and require 'rounding-off' so as not to present an illusion of accuracy (i.e. use '× 1,250' for '× 1,247', and '× 2·5' for '× 2·48'). It is also important to remember that the magnifications in a legend refer to the magnification at the size it will appear in the paper and not that before reduction. Standard magnifications throughout a paper are preferred as this facilitates comparison (e.g. all perithecia × 150 and all spores × 1,000).

When the figures are completed it is advisable to check them carefully against the written description to ensure that they agree precisely, paying particular attention to scales and spore shapes. It is essential that any errors are noted at this stage for while minor slips in the text can often be easily corrected, an error in an illustration will almost certainly mean the expense of a new printer's block. Authors should also be sure that any labels on their illustrations agree with their legends; unfortunately this is not always the case! Many journals require authors to type the legends to their illustrations on separate sheets attached to the end of their manuscript.

It is recommended that the precise material from which a line drawing or photograph has been prepared is indicated in the legend in parentheses after the name of the taxon (e.g. 'FIG. 21. *Mycosphaeropsis callista* (IMI 21230). A, Conidia; B, conidiogenous cells . . .' etc.). Legends should always be sufficiently detailed so that the illustration can be interpreted without reference to the text.

If figures are large and unwieldy they are prone to damage in the post when sent out to referees and so editors often appreciate receiving reduced photographic copies of such figures for this purpose; such reduced versions are, however, almost always of too poor a quality for a printer to use in making blocks.

For general notes on the preparation of illustrations see Hill (1915) and the Royal Society (1965).

Drawings

Newcomers to mycology tend to look at line drawings published by other mycologists and feel that they will never be able to produce figures of such high standards. With practice, however, most students will find that they are able to produce drawings which are quite adequate for publication.

The essential requirement of any drawing is accuracy, for a poor or inaccurate drawing may mislead future workers. Macroscopic (habit) drawings are perhaps the most difficult, but with the aid of feint-blue ruled graph paper and a millimetre ruler satisfactory results are fairly easy to obtain.

Drawings of microscopic features are made with a drawing aid attached to the microscope. The camera lucida (Abbé) drawing device is attached to a vertical ocular tube of the microscope and consists of mirrors which reflect both the image of the object and the drawing paper so that, after adjusting the lighting, they are seen simultaneously. An alternative device now produced for use with some microscopes (e.g. Leitz) is a drawing tube which comes out at right angles to the microscope. Such drawing tubes work on the same principle as the simpler types of camera lucida and although more expensive are easier to use. In both cases all that is necessary to produce an accurate drawing is to draw around the outline of the image with a sharp pencil. Scales may easily be added using an eyepiece micrometer. Camera lucida pencil drawings may be quickly produced with practice and the value of making such sketches in the course of routine examination has already been referred to (p. 19). Particular care is necessary, however, in accurately drawing structures such as ascus tips and the apices of conidiogenous cells.

Pencil drawings themselves are not adequate for publication but need to be gone over in waterproof Indian ink. The most satisfactory way to do this, and at the same time to transfer the drawing to Bristol board, is to use a simply constructed light box; a frosted-glass sheet *c.* 30 × 30 cm inclined at about 25° and illuminated from below. A pencil drawing, if made on normal-weight paper, can be seen easily through a sheet of Bristol board placed over it on such a light box. Redrawing with Indian ink is then simply a matter of tracing round the pencil drawing directly on to the Bristol board. This method also allows one to rearrange spores and transfer drawings from several different sheets of paper on to a single piece of Bristol board without undue fear of introducing inaccuracies.

Attention must always be paid to the thickness of lines used in redrawing, bearing in mind the amount to which the illustration will be reduced. A range of pens giving lines of different thickness (e.g. 0·9 mm, 0·6 mm, 0·3 mm) is necessary; pens designed specifically for this purpose (e.g. Rötring) are easier to use than conventional nib pens.

Merely tracing outlines is not, however, always sufficient for dark coloured structures and cannot provide information as to an object's three-dimensional shape. Careful 'shading' by means of minute ink dots inserted with a fine pen are able to convey impressions of shape and colour. For examples of the use of this method of shading see the figures in Ellis (1971). Some firms now produce self-adhesive shading but this is necessarily artificial and to be avoided in mycological drawings whose aims are realism and accuracy; they are, however, sometimes useful in diagrammatic illustrations.

It is tempting, when drawing complex structures such as a perithecial or pycnidial wall, to draw in hyphae free-hand rather than to draw each one of them accurately. The conscientious student will deplore this approach and prefer either to draw the structure properly or to label it as 'diagrammatic'. Where large parts of a diagram have to be shaded or cross-hatched this is best left to the engraver making the block who can prepare such markings mechanically. In this case clear instructions to the engraver must be inserted in pencil to ensure that no errors are made.

Water-coloured drawings are valuable in the Agaricales but are costly to reproduce; for this reason journals are unlikely to accept them unless they are of a particularly high standard and essential to the work. For excellent recent examples of coloured illustrations in the Agaricales see Reid (1966→) and Romagnesi (1970). In checking the proofs of coloured illustrations it is essential to ensure that the colours have been reproduced accurately as any divergences may prove misleading to later workers.

Slips in drawings are bound to be made by the most practised mycologist but are simply corrected by painting over the error with process white and redrawing.

For further data on the preparation of botanical line illustrations see Blunt (1950), Staniland (1952) and Nissen (1966).

Graphs

Graphs and histograms have only a limited use in most systematic mycological publications but are sometimes used to compare, for example, the size range of spores in different taxa or specimens and to illustrate the results of some methods of numerical taxonomy (see pp. 102–104). Graphs may be drawn in Indian ink directly on to feint blue-squared graph paper which will not show up in the finished printer's block. Only a limited number of curves should be drawn on each to prevent the whole becoming too complicated. All the points used in the plotting of a graph must be indicated so that the reader is able to see immediately how many were used and their divergence from the constructed line. The procedures to be followed in drawing and lettering graphs are similar to those for line drawings. Axes are most satisfactorily drawn at half the thickness of the graph-lines themselves (see the Royal Society, 1965).

Maps

The preparation of distribution maps for publication presents special problems and is treated separately below (pp. 81–89).

Photographs

Black and white (half-tone) photographs are much more expensive to reproduce than line drawings and so are unlikely to be accepted by a journal unless of a very high standard. The mycologist may wish to employ close-up photographs, photomicrographs, or both.

Close-up photographs show the habit of macrolichens extremely well and many lichenologists now always publish photographs of newly described species; they are also necessary to illustrate adequately the appearance of cultures and the effects of microfungi (e.g. characteristic leaf spots).

A 35 mm single reflex camera with a standard 50 mm focal length lens is adequate for close-up work both in the field and in the laboratory. Extension rings (5 mm, 10 mm, 15 mm or 20 mm deep) screwed between the camera and lens will often be found necessary for taking particularly small objects. In the field natural light often provides better results than flash lighting, particularly if colour film is employed. Accurate exposure is the key to a satisfactory result and an exposure meter (cadmium disulphide type) is essential; some single-reflex cameras now have these built into them. In taking a reading, point the meter directly at the specimens from the camera and remember to allow for any extension rings that might have been used. Exposure times with the wide apertures desirable for close up work in natural light are often considerable and will almost always require the use of a robust tripod. For further advice on close-up photography in the field see Campbell (1968).

When taking close-ups in the laboratory adequate illumination may present some difficulties. Two identical small photoflood lamps in reflector stands provide the most satisfactory results when placed at equal distances from the specimen at an angle greater than 45° between the lens axis and the axis of illumination. For deep or irregular specimens shadows may prove difficult to eliminate. Solutions to this problem may be found by (a) suspending the specimens on needles, (b) placing them on a glass plate, or (c) illuminating them from below through a sheet of ground glass. For further information on the close-up photography of specimens see Martinsen (1968).

For microscopic structures a comprehensive microphotographic system (i.e. camera, microscope and exposure meter built into a single unit) is ideal and these are now produced by most of the larger microscope manufacturers. Such systems are unfortunately rather expensive but at least one should be available in all mycological departments. Coloured filters, dark field and phase contrast may be employed to achieve more satisfactory results. A detailed account of photomicrographic techniques is outside the scope of this work. Further general information is given by the Commonwealth Mycological Institute (1968) and Lawson (1972). Tupholme (1961) deals in some detail with colour photomicrography.

Time-lapse and micro-cinematographic techniques (see Burton, 1971; Rose, 1963) have applications in mycology for studying spore development and discharge (see p. 104) but have not been exploited as fully as they might in systematic studies. The applications of transmission and scanning electron microscopy in systematic mycology are discussed separately on pp. 105–106.

Whatever types of photographs are taken care is required in their composition. The part of the field which is to be illustrated particularly is usually most satisfactorily featured centrally and should be focussed upon carefully. Photographers often prefer to take several photographs at slightly different exposures and select the most satisfactory later.

Glossy prints are the most satisfactory for reproduction and only prints of this type should be submitted for publication. General points to be considered in mounting, labelling and submitting illustrations for publication have already been referred to above (pp. 57–58). Where an unmounted set of photographs is required each should be labelled on the back with the author's name, and the plate, figure number and top indicated. Such labelling has to be done lightly with a pencil as ballpoint and heavy pencil marks may reproduce through the print.

In case of difficulty a professional photographer should be consulted. Many university departments and other biological institutions now employ a photographer to assist their staff in this aspect of their work.

Engel (1968) provides a comprehensive introduction to photographic techniques.

Tables

Tables are sometimes valuable in taxonomic papers, particularly for summarizing differences between taxa. These must be typed on separate pages and, like illustrations in general, must have legends which are explanatory without reference to the text. When constructing tables the page size of the journal requires careful consideration. In numerical taxonomic or phytosociological studies large tables which cannot be reduced to the size of a single page may be essential. These are most satisfactorily prepared on an electric typewriter so that printers blocks can be made directly from them and fold-out tables prepared. If the number of such tables is large and the data are also incorporated into other figures or tables it may be better to deposit them in a national library and refer to this in the text.

In tables it is important to distinguish between a negative entry (e.g. character absent) and one for which no data are available (e.g. specimen or organ not studied).

For the format and ruling of tables recent issues of the journal to which the article is to be submitted should be consulted.

MONOGRAPHS AND REVISIONS

A TAXONOMIC monograph is a fully comprehensive account of all that is known of the systematics of a particular group while a revision does not claim to be as exhaustive. Both may be world wide or regionally restricted. Many works entitled 'revisions' by their authors prove in practice to be monographs and are so titled simply because of the author's consciousness of 'failings' in his own work. Both present similar problems and are consequently discussed together here.

Before embarking on any revisionary work the mycologist is advised to consult specialists, recent literature, newsletters and bulletins (see pp. 107–108) to ascertain if any work is proceeding on the group he considers investigating. There is also little point in revising a group for which the existing published works are adequate. To assess the value of previously published revisions the following points should be considered: (1) is the nomenclature correct according to the latest edition of the Code; (2) have many taxa been described since the revision or monograph appeared and have all taxa described prior to its publica- tion been mentioned; (3) do the keys work well and agree with the descriptions; (4) is it based on modern taxonomic concepts; (5) have all taxa been satis- factorily typified and all relevant type specimens examined; and (6) has an adequate amount of material been studied? These are the questions that will be asked of a student's own studies so if he feels that he can improve only slightly on the extant work his talents might be better employed in the study of a less well known group.

The preparation of monographs and revisions is both time consuming and tedious but remains the best way of training the potential taxonomist as it involves critically examining large numbers of specimens, a knowledge of the literature and a clear understanding of both taxonomic concepts and nomen- clature. It is necessary for newcomers to systematic mycology to be supervised by an experienced taxonomist to ensure that their resultant work is as satisfactory as possible.

Preliminary work

The first stage in any work of this type is to compile an index of all the published names which appear to be relevant from appropriate catalogues of names (see p. 108). Each of these should be checked in the original, details of their type specimens (and their location) abstracted, and the status of the names under the Code checked. It is helpful to obtain photocopies of each original description and paste each on a separate sheet in a loose-leaf folder.

A separate card-index of references containing any pertinent information (e.g. descriptions, synonymy) should also be built up. Notes on synonymy, types stated to be missing, chemistry, ecology etc. can also be entered on separate sheets and filed with the original descriptions.

Material

To ensure that the work will be as comprehensive and as useful as possible as much material as time permits should be examined. Usually this will involve borrowing material from other institutions. It will not, of course, be possible to study all herbaria but particular attention should be paid to as many major herbaria as possible and those where other specialists have worked on the group to be studied.

When requesting material on loan the main synonyms should be cited as many herbaria may not have the material filed under the currently accepted name or may not have an adequate system of cross-referencing. Material of an imperfect state, for example, may or may not be filed with the corresponding perfect state. It is often better to visit larger herbaria personally and work on their collections there so that only those specimens requiring more detailed examination need be sent on loan to the worker's own institution. Where the species grows in culture it is also important to obtain subcultures from culture collections (see p. 35) so that their variability can be investigated adequately.

More material than can be coped with at one time should not be obtained on loan as it must be returned before the period of the loan expires unless special permission has been obtained to extend this. Many herbaria use their collections continually for checking identifications and are consequently reluctant to send many of them on loan all at once. Any material borrowed should be treated with the greatest possible care as if it is returned in a damaged condition this may jeopardize future loans to the worker's institution. Type and authentic material must, of course, be treated with particular care and only the minimum of material used; some old type specimens have become indeterminable because later mycologists have, for example, removed all the fruiting bodies over the years (see also p. 31). Any slide made should be returned with the loan or filed in the student's institution's herbarium. Before any material is returned all relevant drawings, measurements, photographs and details of the specimen should be made so as to forestall the student having to request on loan once more material that he has already borrowed and returned.

In addition to general collections of the group all relevant type specimens must be obtained on loan. If no type material is extant all possible steps must be taken to find other authentic material which might provide possible sources of lectotypes (see p. 128). Details of the location of some important mycological collections are given on pp. 179–92.

If the group concerned is readily recognizable in the field (e.g. Agaricales, Xylariaceae, Lichenes) field work should be carried out in as much as possible of the geographical range of the species. Field work often adds much to the understanding of a group. Minor morphological variations, for example, may be found to be correlated with particular substrates, or other environmental factors, and a knowledge of the variability within individual natural populations built up. Workers who limit their studies to herbaria often tend to recognize many more taxa (particularly infraspecific categories) than those familiar with populations in the field. Grove (1937) observed that: 'The old weather-beaten field-naturalists (despite their want of platinum needles) were in many respects nearer to nature, and to the truth, than a great many of their pallid indoor successors of today.'

Delimitation and description

When a representative amount of herbarium material or subcultures is available it should be examined critically and drawings, notes, or chemical studies made as appropriate, and then sorted into the smallest possible entities which the student can recognize as distinct. A careful search for 'intermediates' between his groups should then be made as more material is examined and some will inevitably be found to form part of a continuous range of variation whilst others appear to be clearly distinct. Any possible influence of environmental factors must always be borne in mind at this stage, particularly if little field work has been carried out. The importance of ecological factors in delimiting taxa in the lichens is discussed by Hawksworth (1973b).*

Only when this stage has been reached should any attempt be made to give names to the 'entities' as the monographer who starts by trying to fit specimens

* See also p. 105.

to published taxa is likely to perpetuate errors and overlook doubtful separations. The mycologist will now be in a position to test any existing keys on his 'entities', to begin to decide what ranks to apply to them, and to see to which 'entity' each type specimen belongs. It is often helpful at this stage to construct a key (see pp. 71–75) and allow colleagues or students to try it out; if, using this they can accurately and quickly distinguish between his groups, the revision is progressing satisfactorily.

Full descriptions for each taxon should then be compiled, based on all the available material of the taxon. Some workers find it helpful to have duplicated sheets with all the characters on which data are required listed down one margin with space to fill in colours, measurements etc. next to each. If the monograph is concerned with a particular geographical region the descriptions must be based on material from that region and any variations seen only in specimens from other areas noted separately. Care must be taken to ensure that the descriptions of all taxa are comparable and that comparable items of information are included for each character in each description unless there is some reason why they cannot be. Illustrations and photographs can also be prepared at this stage. Further information on the description of fungi and the preparation of photographs and illustrations is given on pp. 57–62.

If some of his 'entities' do not include the type specimens of previously validly published names he will wish to describe them as new taxa. Under the Code he is required to designate a holotype (see p. 127) for each new taxon in addition to providing Latin descriptions of them (see pp. 56–57, 125, 156). As the holotype is the specimen to which a name is permanently attached and many other workers will wish to examine it in future years it must be carefully selected. The holotype should be representative of the author's concept of his taxon, adequate in amount, and carefully preserved (pp. 31–32). If he has been dealing with cultures one must be dried and preserved as a herbarium specimen to serve as the holotype. Wherever possible duplicates of the holotype (i.e. isotypes) should be distributed to major mycological and lichen herbaria (e.g. BM, BPI, CBS, H, IMI, K, LE, PC, S, UPS, US, W).

When these data have been obtained the student should work out and check his typifications and nomenclature carefully according to the Code (see pp. 124–178), satisfy himself that the taxonomic concepts are consistent between taxa of the same rank (e.g. check that a taxon treated as a variety is less distinct from its species than the other species are between themselves), sort out drawings and photographs which might be suitable for publication, and work out the geographical range and host range of each on the basis of the specimens examined.

Only when he is satisfied that his work is as comprehensive as he can possibly make it should the student begin to think about publication, attach 'determinavit' ('annotation', 'revision') labels (bearing his name, that of the specimen, and the date the label was attached) to the specimens he has received on loan, and return the material he has borrowed.

Scatter diagrams

Where a species or group of species vary in several characters which may be either measured or scored as numerical value character states, this information can often be most satisfactorily displayed graphically in the form of scatter diagrams. If a graph is constructed employing two different measured parameters as its axes each specimen or taxon can then be plotted on it. In addition to the positioning on the axes further information can be incorporated by employing, for example, different shaped symbols and(or) adding dashes (which can be of varying length for measured characters) to indicate different features. In some cases groups of taxa or specimens will be found to form distinct clusters on the diagrams which may lead to the detection of groups not previously apparent.

Although this approach is well established in flowering plant taxonomy (see Davis and Heywood, 1963) it has not so far been employed very widely in the lichens and fungi. For examples of the use of this method in the lichens see Kärenlampi and Pelkonen (1971) and Hawksworth (1972a).

Publication

The monographer hopes that others wishing to determine a specimen of the particular group covered will use his work. This aim is only likely to be achieved, however, if the work is readily available, interesting, comprehensive and lucidly presented. If it is lacking in some respect it may not receive the attention it merits.

A large monograph or revision can usually be divided into three main sections, a 'General Part', 'Special Part' and one or more Indices. The General Part should include an introduction summarizing previous studies on the group, the separation of the group from allied ones and its taxonomic position in the fungi as a whole, comments on the taxonomic value of particular characters, and some indication of the material studied (e.g. a list of herbaria who sent material on loan and the total number of specimens examined).

The Special Part usually comprises a detailed description of the largest taxonomic group concerned (e.g. the family or genus or subgenus) with details of its typification and synonymy, a key to the recognized taxa (see pp. 71–75), and detailed accounts of the taxa dealt with. The taxa may be arranged either alphabetically or according to some taxonomic scheme but where they are not treated alphabetically it is helpful to indicate in the key on which page the account of each starts. Every genus or species entry should include a full list of synonyms (with places of publication and types), a description which is as full as possible (but which need not include characters common to all if these have been included in the description of the next highest taxon) accompanied by a Latin description or diagnosis and the designation of a holotype in the case of new taxa, a list of exsiccatae examined, details of the geographical distribution, ecology and host range, a list of specimens examined (or a list of representative specimens examined or a distribution map if this would be very lengthy), and a discussion on the variability, taxonomy, or nomenclature of the taxon if necessary.

The precise layout of synonymy will in most cases be dictated by the style of the journal, but if the author has freedom of choice one of the arrangements suggested in Figs. 6 and 7 is recommended. In Fig. 6 all names based on the same nomenclatural type are placed together in the same paragraph. Titles of journals are abbreviated according to an accepted system (see pp. 114–115) and the titles of books may also be abbreviated (see pp. 193–199). Care must be taken to recheck all places of publication as errors in page numbers and dates easily arise in the course of retyping. Names which have not been published (i.e.

4. Ascotricha chartarum Berk., Ann. Mag. nat. Hist. ser. 1, **1**: 257 (1838).
    Type: England, Northamptonshire, King's Cliffe, on white printed paper in a deal candle box, *M. J. Berkeley* (K – holotype).
    Chaetomium chartarum (Berk.) Wint., Rabenh. Krypt.-Fl. **1** (2): 157 (1885).
    Non *Chaetomium chartarum* Ehrenb. [Sylv. Beroliens. : 27 (1818)] ex Fr., Syst. Mycol. **3**: 255 (1829).
    Chaetomium berkeleyi Schröt., *in* Cohn, Krypt.-Fl. Schles. **3** (2): 284 (1894). *Nom. nov.* for *C. chartarum* (Berk.) Wint. (1885) non Ehrenb. ex Fr. (1829).
    Chaetomium sphaerospermum Cooke & Ellis, Grevillea **8**: 16 (1879). Type: U.S.A., N.J., Newfield, on bottom of barrel in cellar, *J. B. Ellis* 3174 (K–isotype).
    Chaetomium delicatulum Roum., Rev. Mycol. **7**: 22 (1885). Type: France, Rouen (Seine-Inférieure), développé sur la poudre de scille gâtée, vi. 1884, *A. Malbranche*, Roum., Fung. Gall. Exs., no. 3143 (K, UPS – isotypes).
    Chaetomium zopfii Boul., Rev. Gén. bot. **9**: 25 (1897). Type: Brazil, on bark of *Piscidia erythrina* [no further data] (not traced).
    Ascotricha zopfii (Boul.) Peyronel, Annls mycol. **12**: 464 (1914).
    Ascotricha chartarum var. *orientalis* Castell. & Jacon., J. Trop. Medic. Hyg. **37**: 362 (1934); as ' *Ascothrica* '. Type: China, Prov. Shensi [received from A. Castellani] (CBS 117.35 – lectotype).
    Myxotrichum murorum Kunze, *in* Kunze & Schmidt, Mykol. Hefte **2**: 110 (1823). Type: (L 90.OH.910, 255-336 – holotype; not seen); *stat. conid.* (see Hughes, 1968).
    Sporodiniopsis murorum Höhnel ex Lindau, Rabenh. Krypt.-Fl. **1** (8): 268 (1905); as ' (Kze.) '; *stat. conid.*
    Dicyma ampullifera Boul., Rev. Gén. bot. **9**: 20 (1897). Type: Brazil, isol. ex bark of *Piscidia erythrina* [no further data] (not traced); *stat. conid.*
    Dicyma chartarum Sacc., Syll. Fung. **18**: 570 (1906). Type: Zopf & Sydow, Mycoth. March. no. 69 (K, K ex BM, UPS – isotypes); *stat. conid.*
    The conidial state was also given a herbarium name by M. C. Cooke (in K) but was not validly published by him.

Fig. 6. Layout of synonymy (after Hawksworth, 1971*b*).

herbarium names) and which are not being validated should be omitted (see Fig. 6). If published names are invalid or illegitimate it is helpful to indicate this by appending '*nom. illegit.* (Art. 64)' or '*nom. inval.* (Art. 36)' etc., after the citation. Where a list of synonyms of the perfect state of a fungus includes those of the imperfect state it is recommended that '*stat. conid.*' or '*stat. pycnid.*' be added after the citation of the imperfect state (see glossary). Details of the type specimens should ideally be included next to the name of which they are the type and the herbarium where they are located, and the nature of the type (i.e. lecto-type, holotype, neotype etc.; see pp. 127–128) indicated. If type material or the reference given to a synonym has not been seen or checked, respectively, this can be indicated by adding '*not seen*' or '*n. v.*' in parentheses.

A

11. Alectoria smithii DR.

Ark. Bot. **20A**(11): 15 (1926); type: China, Prov. Sze-shu'an, inter Tsago-gomba et Tamba, in *Juniperus, Picea* et *Rhododendron*, alt. 4000 m, 2 October 1922, *H. Smith* 5025b (UPS – lectotype). – *Alectoria berengeriana* var. *smithii* (DR.) Gyeln., *Annls Mus. Nat. Hungar., Bot.* **29**: 6 (1935). – *Bryopogon berengerianus* var. *smithii* (DR.) Gyeln., *Feddes Repert.* **38**: 233 (1935). – *Alectoria bicolor* subsp. *smithii* (DR.) Räs., *Annls Bot. Soc. zool.-bot. fenn. Vanamo* **12**(1): 32 (1939).

Alectoria bicolor var. *berengeriana* Massal. ex Stiz., *Annls Naturhist. Hofmus. Wien* **7**: 127 (1892); type: Italy, Prov. Vicentinae, Cadore, *Berenger*, Anzi, Lich. R. Venet. Exs. no 17. (ZT – lectotype; BM, BM ex K – isolectotypes). – *Alectoria bicolor* subsp. *berengeriana* (Massal. ex Stiz.) Suza, *Věstn. Krat. České Spol. Nauk, Třída mat.-přirod.* **1933**(9): 15 (1934). – *Alectoria berengeriana* (Massal. ex Stiz.) Gyeln., *Nyt Mag. Naturv.* **70**: 61 (1932). – *Bryopogon berengerianus* (Massal. ex Stiz.) Gyeln., *Feddes Repert.* **38**: 232 (1935).

Bryopogon divergens var. *rigidum* Hazsl., *Magy. Birod. Zuzmó-Fl.*: 28 (1884); type: Hungary, Csarna gura, *F. Hazslinskyi* (BP no. 33.937 – holotype). – *Cornicularia divergens* f. *rigida* (Hazsl.) Zahlbr., *Cat. Lich. Univ.* **6**: 414 (1920). – *Bryopogon berengerianus* var. *rigidum* (Hazsl.) Gyeln., *Feddes Repert.* **38**: 232 (1935). – *Alectoria csarnagurensis* Gyeln., *Nyt Mag. Naturv.* **70**: 44 (1932).

Alectoria bicolor f. *major* Hav., *Lich. Norv. Occ. Exs.* no. 30 (1913), nom. inval. (Art. 32).

Alectoria csarnagurensis f. *grisea* Gyeln. & Fóriss, *Magy. Bot. Lapok* **30**: 53 (1931); type: Hungary, Com. Hunyad, Kudsiri Havasok, Cultul Marului, alt. 1000 m, in cortice *Fagi*, 28 July 1912, *E. Fóriss* 2246 (BP no. 33.938 – holotype). – *Bryopogon berengerianus* var. *rigidus* f. *grisea* (Gyeln. & Fóriss) Gyeln., *Feddes Repert.* **38**: 233 (1935).

Alectoria sandwicensis Magnusson, *Ark. Bot.* **32A**(2): 1 (1945); type: Hawaii, E. Maui, Haleakala, outside of crater along road, alt. 8000 ft, 5 August 1938, *O. Selling* 5890 (S – holotype).

The holotype collection of *A. smithii* was destroyed and the selection of the lectotype is discussed by Jørgensen and Ryvarden (1970).

B

9. **Myrothecium verrucaria** (Alb. & Schw.) Ditm. ex Fr., Syst. mycol. **3**:217, 1829
≡ *Peziza verrucaria* Alb. & Schw., Consp. Fung. p. 340, 1805
= *Gliocladium fimbriatum* Gilman & Abbott, *Iowa St. Coll. J. Sci.* **1** (3) 304, 1927
= *Metarhizium glutinosum* Pope, *Mycologia* **36**:346, 1944
= *Starkeyomyces koorchalomoides* Agnihothrudu, *J. Indian bot. Soc.* **35**:41, 1956.

Fig. 7. Layout of synonymy; A (after Hawksworth, 1972a), B (after Tulloch, 1972).

Any published names which are not included in the synonymy of any of the taxa treated can be included under a list of 'excluded names'. An index to all the names treated is valuable as it makes the checking of the identities of names no longer accepted much easier. If the group is one which shows pronounced host specificity a separate host index is also valuable.

Citations of specimens examined · must always be given as precisely as possible. Various methods of citing herbarium specimens have been employed by mycologists and lichenologists; they are usually arranged according to their herbarium folders, hosts, substrates, country, province, county, district or grid reference (e.g. Fig. 8–9). For each specimen all pertinent data on the herbarium

A

Sumatra. Oostkust v. Sumatra, Pisopiso (at Lake Toba), alt. c. 1200 m, 1926 Palm n. 25 a (Magn.). Enggano Isl. (off SW. Sumatra), behind Ekinoia, 1936 Lütjeharms n. 4949 e (L). — Java. Batavia, Mt Megamendoeng, Telaga Warna, alt. c. 1400 m, 1894 Schiffner n. 3349 (UPS). Salak, 1898 Nyman n. 45 a (UPS). Preanger, Tjibodas, 1895 Massart n. 1339 (PC); same loc., alt. c. 1500 m, 1898 Nyman n. 3 a, 6 b, 49 c, 87 d (all in UPS). — Philippines. Luzon, Camarines, Adiagnao, 1908 Robinson n. 6392 (TUR), without n. (TUR). Polillo, 1909 Robinson n. (TUR, UPS). — Moluccas. Amboina, 1913 Robinson n. 2418 (L, UPS). — New Guinea. NE. New Guinea, Stephansort, 1899 Nyman n. 100 b (UPS).

Australia. N. Queensland, Atherton Tableland, Gadgarra State Forest, mixed rain-forest, 1927 G. E. Du Rietz n. 4294:a (Calamus?; S). Kuranda, The Maze, mixed rain-forest on shrubs, 1927 G. E. Du Rietz n. 4267:2 a (S).

B

Other specimens examined: England, Notts., 1870, *M. C. Cooke* (K).—England, Banstead Woods, cardboard covering plaster dumped in hedge, 4. i. 1948, *S. J. Hughes* (IMI 21150d).— England, Holloway, on wallpaper, 1887, *M. C. Cooke* (K).—England, *sine loc.*, on paper, *M. J. Berkeley* (UPS herb. E. Fries).—England, *sine loc.*, bookback carcase, comm. 2. vi. 1949, *F. D. Armitage* (IMI 35213).—England, *sine loc.*, on lining paper of book comm. 18. vii. 1949, *F. D. Armitage* (IMI 35732).—India, comm. University of Allahabad, on leaves of *Sanseviera macrophylla*, comm. 3. xii. 1962, comm. *K. S. Bilgrami* A-2 (IMI 96853).—India, Jabalpur, isol. ex grassland soil, comm. 28. iv. 1971, *P. D. Agarwal* 51 (IMI 157287).—Italy, Vittorio (Treviso), in sparta palacea putri, viii. 1897, [coll. uncertain] Saccardo, Myc. Ital. no. 63 (K ex BM).—New Zealand, Levin, N.Z. Department of Agriculture Horticultural Research Laboratory, isol. ex *Gladiolus* bulb [intercepted from British Isles], 11. vii. 1966, *R. J. Bishop* 833 c (IMI 123231).—Tanzania, Mlingauo, Sisal Research Station, isol. ex sisal twine (*Agave sisalana*), 19. xii. 1966, *J. F. Wiek* 110 (IMI 124033).—U.S.A., N. Hamps., Hanover, on barrel in cellar, 10. vii. 1912, *A. H. Chivers* (K).—isol. ex mouldy linoleum, 1930, *H. Klebahn* (CBS 107.30).—isol. ex waste paper, 1952, *A. Saccas* 420 (CBS 110.52).—isol. ex wine bottle cork, *G. L. Hennebert* (CBS 902.69).—[no data] (CBS 104.25).

C

SPECIMENS EXAMINED

Harknessia uromycoides folder in Herb. LPS.
On *Eucalyptus globulus*, Buenos Aires, Flores, Argentina, v. 1880, C. Spegazzini, LPS 12061, type of *H. uromycoides* (148593); LPS 11822 (148592).

Harknessia uromycoides folder in Herb. K.
On *Eucalyptus odoratus*, Piedmont, April, no. 3079, type of *H. longipes* (146761); on *Eucalyptus globulus*, Berkeley, Calif., U.S.A., ix. 1893, W. C. Blasdale, Rabenhorst-Winter-Pazschke Fungi europaei 4076 (146777); ii. 1931, E. B. Copeland & H. E. Parks, California Fungi 407 (146776); Coimbra, Portugal, i. 1887, A. Moller, C. Roumeguère Fungi Gallici Exsiccati 4070 (146772); on *Eucalyptus* sp., W. Australia, vii. 1915, F. Stoward (146774b); Perth, W. Australia, viii. 1915, F. Stoward 237 = 37 (146775); Calif., U.S.A., H. W. Harkness, Ell. & Ev. North American Fungi 2nd ser. 1649 (146773).

Harknessia eucalypti folder in Herb. K.
On *Eucalyptus globulus*, Stanford Univ., Calif., U.S.A., i. 1903, E. B. Copeland, Pacific slope fungi 2723 (146780); Berkeley, Calif., U.S.A., vi. 1893, W. C. Blasdale, Ell. & Ev. Fungi Columbiani 80 (146783); Compton, Calif., U.S.A., iv. 1897, A. J. McClatchie, Ell. & Ev. Fungi Columbiani 1149 (146781).

Harknessia uromycoides folder in Herb. IMI.
On *Eucalyptus globulus*, Coimbra, Portugal, i. 1883, A. Moller, Rabenhorst-Winter Fungi europaei 2987, type of *H. molleriana* (16588); Ocean Rd, Lorve, Australia, vii. 1966, G. Beaton (121052); v. 1965, K. & G. Beaton GB5 (115812); Calif. Univ., Calif., U.S.A., i. 1965. I. Tavares & J. Bock 1679 (142230); on *Eucalyptus gomphocephala*, Younghusbands Penin-sula, The Coorong, Australia, x. 1967, Beauglehole, Williams & Beaton (130785); on *Eucalyptus* sp., Santa Clara Co., Calif., U.S.A., 1901, C. F. Baker, type of *Sphaeropsis stictoides* (6109).

Fig. 8. Methods of citing specimens examined; A, by country with a new paragraph for each subcontinent and continent (after Santesson, 1952); B, by country with a dash separating each collection (after Hawksworth, 1971b); C, by herbarium folders examined (after Sutton, 1971).

packets should be included, and the herbarium in which it is located indicated (see pp. 187–192 for some herbarium abbreviations). The most appropriate method for a particular revision may vary but the arrangement in Fig. 8A is the most satisfactory for general use. In citing specimens localities should be verified from appropriate atlases and gazetteers (see p. 86–89) and their spelling checked. Where a place name has been misspelt or a place has changed its name (e.g. in USSR) the currently accepted spelling and name can be placed in square brackets

A   *Localities:* **V.C. 7,** North Wilts: Marlborough, Savernake Forest, near King Oak, on old *Quercus* by side of path, 19 September 1968, *F. Rose* (BM); also in same area, nat. grid 41,255,659, 41,214,675, 41,224,657, and 41,212,672, all on old *Quercus,* 15 February 1969, *P. W. James & F. Rose* (BM). **V.C. 11,** South Hants: New Forest, Romsey, Bramshaw Wood, nat. grid 41,263,167, on *Quercus,* 6 March 1968, *P. W. James & F. Rose* (BM). **V.C. 13,** West Sussex: South Harting, Upark, on isolated *Quercus* in pasture, March 1971, *B. J. Coppins* (BM). **V.C. 76,** Westmorland: near Kendal, Levens Park, nat. grid 35,49–,85–, on a single *Quercus robur,* 5 June 1969, *B. J. Coppins, D. L. Hawksworth & F. Rose* (BM, IMI).

B   *British specimens examined*
        –/– Highlands of Scotland, November 1778, Herb. Menzies (E); *sine loc.,* December 1806, sent to W. Turner (BM); *sine loc.,* 18 – , Herb. H. Davies (BM). 20/58 N. Devon: Black Tor Beare, on *Quercus,* alt. 390 m, 11 August 1969, *F. Rose & T. D. V. Swinscow* (DLH no. 2124, FR, TDVS); 5 September 1971, *D. L. Hawksworth* (DLH no. 2704). 20/67 S. Devon: Wistman's Wood, 7 August 1939, *T. Stephenson & W. R. Sherrin* (BM, BM ex K); on *Quercus,* 7 April 1969, *D. L. Hawksworth* (DLH no. 1852, FR). 23/53 Merioneth: Moel Tfridd, above Harlech, January 1923, *D. A. Jones* (BM ex K). 23/62 Merioneth: lake above Llyn Bodlyn, 18 –, Herb. W. Borrer (K – Borrer); Llyn Bodlyn, 1877, *T. Salwey* (BM). 23/63 Merioneth: nr Cwm Maws, nr Llanbedr, boulder in grassland above stream, September 1959, *S. A. Manning* (SAM). 23/65 Caernarvon: New Glyder, 1788, Herb. H. Davies (BM). 27/53 Perth: Meall Ghaordie, 18 – , *Brodie* (E). 27/79 Stirling: Thalsfield, 1782, *Buchanan* (LINN-Smith coll.).

Fig. 9. Methods of citing specimens examined; A, by vice-county number (after James, 1971); B, by 10 km square grid reference (after Hawksworth, 1972*a*).

following that on the herbarium packets. Considerable care is needed in reading handwritten labels in languages other than the author's own (particularly in the case of pre-1900 collectors). Dates should never be given in the form 1.8.49; this example could refer to 1 Aug. 1849, 1 Aug. 1949, 8 Jan. 1849 or 8 Jan. 1949. Often the year of collection (or isolation) is adequate but fuller citations may be helpful in herbaria where packets are unnumbered as a collector may have collected many specimens from one site in a particular year. Collectors' numbers should be included where any are given on the packets. The citation of specimens in floristic studies is discussed further on p. 80.

    Most journals require a summary or abstract of longer articles but even if they do not stipulate this in the case of lengthy monographs and revisions these are important. The summary should include such information as the scope of the study, the number of species accepted and name changes introduced, and should list all new taxa and combinations. A good summary facilitates the task of the compilers of abstracting periodicals and reduces the chance of new taxa being missed (e.g. by the *Index of Fungi*).

Before writing up work in a form which will be submitted to an editor for publication, it is advisable to study the layout of as many published monographs by more experienced workers as possible as these may yield ideas for the improvement of the manuscript.

When the mycologist has his manuscript in the best form in which he can present his data, feels he can improve on it no further, and has carefully checked his keys, descriptions and references, it is advisable to submit it to specialists also familiar with the group for criticism before sending it to an editor. It is better that any slips or omissions should be brought to light at this stage before the paper has been published.

General guidelines on the preparation of manuscripts and their publication are given on pp. 116–123. Davis and Heywood (1963) and Leenhouts (1968) provide further useful information on the preparation of monographs and revisions.

KEYS

KEYS are means of identifying taxa and form important parts of revisions, monographs and larger floras. The ability to construct a workable key to a group of taxa is also a test of whether its taxonomy is sound. When key construction proves difficult this often indicates that further work is necessary and may also serve to draw attention to taxa which may be more closely allied than originally thought. Adequate keys do not exist for many fungi and consequently taxonomists frequently have to construct their own to facilitate the naming of their material.

Keys may be either (a) *natural*, reflecting taxonomic relationships, or (b) *artificial*, providing a rapid means of identification regardless of taxonomic affinities. In some instances the characters reflecting taxonomic relationships are those most suited to accurate determination but in many they are not. As the aim of an artificial key is a correct identification the characters used must always be those which are unlikely to be misinterpreted and easy to observe. It is clearly easier to observe gross morphological characters, colour or substrate, than to study ascus structure, spore ornamentation or chemical components. Maximum accuracy in a key is obtained by using characters which are readily observable under all conditions in which the key is likely to be used, tests whose comparison is simple, and decisive characters which are well separated. The possibility of error in using a key almost always increases with its length if there is any possibility of the user taking wrong decisions.

The most widely used keys are of the artificial analytical type in which contrasting statements (*leads*) are compared. Two leads are used at each point in *dichotomous* keys and several in *polychotomous* keys. Dichotomous keys appear to be the most satisfactory because the fewer the number of alternatives the user has to consider the less the probability of error. Dichotomous keys may be presented in one of two forms, either (a) *bracketed* (or parallel) in which the two leads of each contrasting pair (*couplet*) are printed beneath one another (Fig. 11)

or (b) *indented* (or yoked) in which each lead is followed by all its subordinate couplets so that the contrasting leads of a particular couplet are spacially separated (Fig. 10). Both these types of dichotomous key are widely used in mycology and both have their own advantages: bracketed keys are easier to use as the leads of each couplet appear next to each other, whilst indented keys group species

1a. Thallus orbicular, with lobes at the circumference. On rock.
    2a. Spores 8 in the ascus. Centre of thallus becoming a mass of minute scaly isidia. Common on limestone in Lowland Britain.
                                C. medians (Nyl.)A.L.Sm.
    2b. Spores many in the ascus. Centre of thallus of convex granules. Arctic species on acid sea cliffs used by colonies of birds for nesting. Rare; recorded only from Kincardineshire.
                  C. arctica (Körb.)R.Sant.(*C. crenulata* (Wahlenb.)Zahlbr.)

1b. Thallus crustaceous or evanescent. On rock, bark wood and earth.
    3a. Thallus with numerous spherical granules
        4a. Granules always forming areolae. Granules 0·05–0·3 mm diameter. On acid rock.   *C. coralliza* (Nyl.)Magnusson
        4b. Granules continuous to scattered, rarely forming areolae. Granules 0·01–0·05(–0·1) mm diameter. On bark and wood.
                              C. xanthostigma (Ach.)Lett.
    3b. Thallus of convex subsquamulose areolae, convex granules, or evanescent.
        5a. Spores 8 in the ascus. Thallus poorly developed, of scattered convex granules to evanescent, or of a black prothallus. Apothecia abundant. Calcareous rock; abundant on concrete in urban areas, local elsewhere.   *C. aurella* (Hoffm.)Zahlbr.
        5b. Spores (12–)16–32 in the ascus. Thallus usually well developed, of convex subsquamulose areolae or of scattered convex granules. Common on acid and calcareous rocks, bark, wood and earth.
                        C. vitellina (Hoffm.)Müll.Arg.

Fig. 10. Indented key to the British species of *Candelariella* (after Laundon, 1970).

1. Thallus orbicular, with lobes at the circumference. On rock. 2
   Thallus crustaceous or evanescent. On rock, bark, wood and earth............ 3

2. Spores 8 in the ascus. Centre of thallus becoming a mass of minute scaly isidia. Common on limestone in Lowland Britain .... *C. medians* (Nyl.)A.L.Sm.
   Spores many in the ascus. Centre of thallus of convex granules. Arctic species on acid sea cliffs used by colonies of birds for nesting. Rare; recorded only from Kincardineshire. .. *C. arctica* (Körb.)R.Sant. (*C. crenulata* (Wahlenb.)Zahlbr.)

3. Thallus with numerous spherical granules 4
   Thallus of convex subsquamulose areolae, convex granules, or evanescent...... 5

4. Granules always forming areolae. Granules 0·05–0·3 mm diameter. On acid rock .... *C. coralliza* (Nyl.)Magnusson
   Granules continuous to scattered, rarely forming areolae. Granules 0·01–0·05(–0·1) mm diameter. On bark and wood .... *C. xanthostigma* (Ach.)Lett.

5. Spores 8 in the ascus. Thallus poorly developed, of scattered convex granules to evanescent, or of a black prothallus. Apothecia abundant. Calcareous rock; abundant on concrete in urban areas, local elsewhere. *C. aurella* (Hoffm.)Zahlbr.
   Spores (12–)16–32 in the ascus. Thallus usually well developed, of convex subsquamulose areolae or of scattered convex granules. Common on acid and calcareous rocks, bark, wood and earth........ *C. vitellina* (Hoffm.)Müll.Arg.

Fig. 11. Same key as in Fig. 10 arranged as a bracketed key.

according to their similarity in the lead characters employed. Long indented keys should, however, be avoided as in them it is often necessary to pass over several pages to find both leads of a particular couplet, and they waste space when printed. Each couplet is generally assigned a number, and each lead may be labelled 'a.' or 'b.' to make them clear. Some authors have used symbols or letters instead of numerals but these should be avoided to minimize the possibility of confusion. In bracketed keys it is also helpful to indicate in parentheses after the number of a couplet that of the couplet which leads to it as this makes back-checking much easier. Dichotomous indented keys may also be presented as *solid* keys (see Leenhouts, 1966a) but this type of format is difficult to follow and no longer used.

Construction

The first step in constructing any key is to decide what taxa are to be included. Each may then be entered in a table and scored for particular characters. From such a table the characters to be used in each couplet can be selected readily. The simplest couplets should be used first as mistakes are likely to be more serious earlier in a key than later. It is often also convenient to key out the most commonly occurring taxa early on in the key although this means that the user will become less aware of the characters of rarer species. Each couplet ideally need only contrast two states of a single *decisive* character but it may often be useful to add *accessory* characters in support of these to reduce the possibility of error and to allow, for example, for parts of the specimen which may be missing. Leads often contain statements such as '. . . or if . . . then . . .' and while these may be unavoidable in some cases steps should be taken to ensure that they are minimized. Each lead of a couplet should clearly contrast states of the same character and phrases like 'not as above' deleted and the appropriate character states included. Where a particular taxon is keyed out it is also often helpful to present data on several accessory characters to make the identification certain; information on geographical distribution may also be included here but should not be employed in leads as species may be found to occur outside their generally accepted geographical ranges. In some cases it proves necessary to key out the same taxon in more than one place to avoid ambiguity.

Considerable thought should be given to the couplet characters as their clarity is essential for accurate identification. Magnusson's (1940) use of species 'better known' and species 'incompletely known' as leads of a single couplet clearly leaves much to be desired! The author should also bear in mind that many of the students who will use his key will not be familiar with the group concerned and consequently the language of the leads should be as simple as possible. Where many technical terms prove necessary a glossary and(or) diagrams illustrating them are clearly desirable.

When a draft key has been prepared the author should allow colleagues to use it to ascertain if it works satisfactorily. A couplet which appears quite clear to the specialist may be found to confuse the non-specialist and lead to error. The final key will clearly be more satisfactory if such points are discussed and dealt with before the key appears in print.

In very long keys it is often valuable to provide summaries of characters in different face type at various points (see e.g. Ames, 1963; Duncan, 1970; Thomson, 1968) so that the user is more likely to realize he has made a slip in its use. Blank lines printed between couplets in artificial bracketed keys should be included where possible as the user is then less likely to glance at a number above or below the correct one. Where taxa are keyed out the number which they are given in the work should be indicated or the page number on which it is dealt with placed in parentheses after the name. In the same group two or more keys may be desirable; for example one based on fertile and one on sterile material.

For further information on types of keys and their construction see Metcalf (1954), Stearn (1956), Leenhouts (1966a, 1966b, 1968), Davis and Heywood (1963), and Pankhurst (1971). Osborne (1963) presents an interesting account of the mathematical theory of dichotomous keys.

Synoptical keys

Synoptical or comparative keys are a form of artificial key in which each character in turn is listed with its states and the taxa showing the particular states listed below each character state. Such keys are useful as a step in the construction of analytical keys and for identification of taxa separated by combinations of characters. Synoptical keys have rarely been used in mycology (see Korf, 1972) and while they are useful as summaries of data they perhaps are less suited to rapid identification than analytical bracketed and indented keys. Leenhouts (1966b) gives a detailed account of this type of key, its construction and presentation.

Computer generated keys

Computer generated analytical keys have several theoretical advantages over manually constructed ones. They minimize the labour of construction, will be as accurate as the data, may be modified with little effort, permit particular taxa to be removed or added easily, and allow different weighting of characters to form keys suited to different needs (e.g. sterile material, cultural or chemical characters). Two main methods of generating keys by computer are now available. That of Pankhurst (1970, 1971) and Pankhurst and Walters (1971) is perhaps the more satisfactory, but that of Morse (1968), Shelter et al. (1971) and Hall (1970) has also been employed.

Computer generated keys are clearly more valuable to mycologists wishing to construct keys to large numbers of taxa than to ones dealing with very few. Authors dealing with large numbers of taxa should investigate the possibilities of using such methods further as they will tend to eliminate the possibility of errors due to oversights of characters (which all too frequently occur in lengthy keys).

A novel computer graphics method of determining fungi by image retrieval has been described by Kendrick (1972).

Punched card keys

Punched card systems may be treated as a special type of key which is particularly well suited to large groups where there are many characters separating species, as the user can select all those taxa showing particular characters or combinations of characters rapidly and accurately. The character and character states in the group are first all allotted unique numbers. A punched card (i.e. a card with numbered holes around its margins) is then made out for each taxon and the characters it possesses indicated by clipping out a notch for the number concerned (Fig. 12). When all the cards are aligned, a steel needle is passed

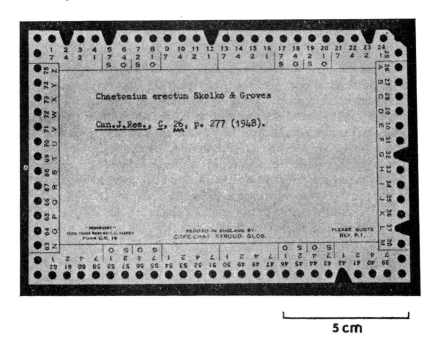

5 cm

Fig. 12. A punched card for *Chaetomium erectum* prepared by Dr C. Booth as part of a punched card key for the identification of *Chaetomium* species at the CMI.

through the hole of a particular character present in the specimen to be named. The cards are then lifted and shaken, so that those of all taxa showing that character state fall out on to the table. The process is repeated until only one card remains; that for the specimen being studied. As the user can select what characters he wishes in any convenient order such systems are sometimes referred to as 'multiple-choice keys'.

The precise type of card to be employed depends on the number of character states (i.e. holes). A good example of this type of key is that designed by the Forest Products Research Laboratory (1960) for the identification of hardwoods. Keys of this type have been used occasionally for both fungi (e.g. Nobles, 1965; Olá'h, 1970) and lichens (e.g. Martin, 1968) and perhaps deserve to be more widely employed. A detailed account of available punched card techniques has been compiled by Casey *et al.* (1958).

FLORAS

A COMPREHENSIVE flora is the key to the understanding and literature of the plants of a particular geographical region just as a systematic monograph is the key to that of a taxonomic group. Monographs and floras complement one another and are equally essential, but whilst monographs must remain primarily the domain of the specialist the compilation of floras can be undertaken by amateurs prepared to devote the necessary time to their preparation (see p. 12).

Before starting on any floristic work the geographical area and the groups of fungi and lichens to be included must be decided upon. If the area chosen is too large the final flora may be rather superficial and overlook many species whereas if it is smaller fewer species are likely to be missed. Floras may be prepared for several countries, individual countries, divisions of countries (e.g. states, counties), parishes or smaller areas such as nature reserves (or even gardens). The whole of the fungi and lichens may be included or only a relatively small taxonomic group (e.g. Ascomycotina, lichens, Agaricales).

Mycological floras often include records of species isolated from, for example, soil, and so the mycologist must decide whether he is going to make isolations or not (see pp. 33–34 for references to techniques). As flowering plant taxonomists do not usually isolate and germinate seeds the place of isolations in mycological floras is perhaps a matter for debate.

Collecting and compiling the data

The first stage in preparing any flora is to ascertain what studies have previously been conducted in the area. The principal fungus and lichen floras for whole countries are listed in Ainsworth (1971) but others should be searched for and as complete a bibliography as possible compiled. Fungal and lichen records are often included in nineteenth century floras dealing primarily with vascular plants and so these must also be checked (for world list see Blake and Atwood, 1942, 1961; for British list see Simpson, 1960). Local museums and abstracting journals also require consultation.

All collections made in the past were not, of course, published and consequently herbaria likely to or known to contain material from the region must also be examined. Unpublished manuscripts and field note books are also valuable sources of information. As far as is possible all specimens supporting published records should be traced and their identification checked.

The author should attempt to study the area in as much detail as is practicable himself and make collections and lists from as many localities at as many seasons as is possible. A first trip will usually yield the commoner species but later careful visits often lead to the most interesting records. Comprehensive floristic studies are essentially long-term projects and cannot be hurried. Field work extending over 3–5 years is to be regarded as minimal.

It is not always possible to retain every single specimen but care must be taken to ensure that representative ('voucher') material of each species is collected and ultimately deposited in one of the main herbaria (see pp. 187–192). If this is not done future workers will have no means of confirming particular reports. Dennis (1968) considered that 'Lists of records that cannot be verified are mere waste paper'.

The compiler of a flora cannot expect to be an expert on all groups of fungi and lichens and should not be afraid to send his collections to appropriate experts. This does not mean that every specimen collected should be sent. Specimens should always be examined and determined as accurately as possible before submitting them to an expert for checking. When specimens are returned they should be studied and learnt so that the same species need not be sent again. Specialists soon tire of helping those who are not prepared to attempt to identify material for themselves. Notes on the collection, preservation and labelling of material have been given on pp. 13–19 and 29–30.

Sorting out the data often presents difficulties. The most satisfactory method is to use either a card-index file or a loose-leaf book so that a single card or page can be devoted to each taxon. While it is possible to include details of every single collection and record this becomes exceedingly laborious for very common species. Most recent British floras use the 10 km square units of the Ordnance Survey National Grid for recording and this has been found to be particularly satisfactory for indicating the distribution of commoner species. An example of a species card used in the compilation of *The lichen flora of Derbyshire* (Hawksworth, 1969a) is shown in Fig. 13. In this case the species concerned proved to be not uncommon and when this had been established details of further square records were not entered on the card and the square numbers crossed off (details of these can be found in field note-books which should also be preserved). The type of published entry which can be prepared from cards of this type is shown on p. 80.

When to cease compilation and start to consider publication is a difficult decision to make. This may be governed by practical considerations such as the compiler moving away from an area or being unable to visit it again, but should ideally be when the author considers his work to be reasonably comprehensive, at least as regards the total list of species. The compiler will eventually find that the numbers of 'new' records resulting from field excursions begin to decline; whereas his first trips might have led to 30 or more additions later ones might yield none or only one or two. Collectors invariably have 'blind spots' for particular groups, however, and so it is advisable to persuade specialists to visit the area being investigated. If they too find little new then the time to prepare for publication is clearly approaching. From the study of Dennis (1973) it is evident that the proportions of the major groups of non-lichenized fungi to be expected in British floras are likely to be Basidiomycotina 47–57%, Ascomycotina 24–27%, Deuteromycotina 12–20%, 'Phycomycetes' 2–6% and Myxomycota 1–5%; local lists of British fungi which are markedly at variance with these values are likely to be based on biased or inadequate collecting. Compilers should also be suspicious of the thoroughness of their work if they have not rediscovered many species reliably reported by previous workers, and the reasons for the absence of these is not clear.

A

90	00	10				
99	09	19	29			
98	08	18	28	38	48	58
97	07	17	27	37	47	57
96	06	16	26	36	46	56
		15	25	35	45	
04	14	24	34	44		
03	13	23	33	43		
		12	22	32	42	52
		11	21	31	41	
		10	20	30		

Acarospora fuscata (Nyl.) Arnold

FIRST RECORD:

Present work.

PUBLISHED RECORDS: ————

HABITAT:

Millstone grit rocks, walls and gravestones only.

B

G.R.	LOC.	DATE	G.R.	LOC.	DATE
43/14	gravestone, Snelston ch.	15 vii 1967	43/26	Rowtor Rocks	15 vii 1967
43/21	Overseal church	9 vi 1967	43/27	Froggat Edge	15 vii 1967
43/16	Monyash churchyard	2 vi 1967	43/18	Hope parish church	15 ix 1967
43/26	Stanton Moor, N.Wallace (BM).	20 vii 1966	43/08	Kinder Downfall	16 ix 1967
43/46	Hardwick Hall	17 iv 1967	43/33	village green, Horsley	6 iv 1968
43/35	Nether Heage (DLH 399)	1966	43/27	Upper Padley Wood	15 iv 1968
43/32	Ticknall village	1966	43/26	Stanton Park, ?Bloxam (MANCH, sub A. macrospora)	18 —
43/26	Beely Lodge, Chatsworth	4 iv 1967	43/26	Youlgreave, 18—, ?Bohler (MANCH, sub A. macrospora).	18 —
43/37	Hare Edge (DLH 697)	14 iv 1967			
43/24	Kedleston church	9 vii 1967			
43/22	Radbourne church	10 vii 1967			
43/23	Willington church	10 vii 1967			

Fig. 13. Record card for *Acarospora fuscata* used in the preparation of *The lichen flora of Derbyshire* (Hawksworth, 1969a); A, front; B, reverse.

Computer produced floras

The advent of computerization has enabled much larger amounts of floristic information to be compiled more readily than was previously possible. The taxa recorded or likely to be recorded are entered on record cards similar to or the same as those used in Mapping Schemes (see pp. 85–88). Separate cards are then completed for each site and(or) habitat visited and the data stored on magnetic computer tape. Programs can then be constructed which enable print-outs of all the records for individual species and(or) maps to be obtained. Because no manual transcription of the data from lists to cards is involved the likelihood of errors is reduced, assuming the input has been properly checked. The computer will not forget the odd list records which are often overlooked by flora compilers and are able to save the flora writer very considerable amounts of time. No computer aided fungal or lichen flora appears to have been published so far but several are in preparation. For an example of a vascular plant and bryophyte flora produced by computer methods see Cadbury *et al.* (1971).

Publication

The final content and presentation adopted in published floras varies considerably. The minimum requirements are an introduction and an annotated species list. The introductory sections should include the precise delimitation of the area (preferably with a map); brief notes on its geology, topography, climate and vegetation; details of previous mycological work in the area; a list of the herbaria examined; and information on the extent of the author's own field work and the whereabouts of his collections (and record-cards and field note books where appropriate).

It is also useful to include a section summarizing the species and communities characteristic of particular hosts or substrates (using phytosociological methods where possible; see pp. 89–95), to point out any distinctive geographical elements which may be present (with the aid of distribution maps where appropriate) and to compare the flora with that of neighbouring regions. Some lichen authors provide a richness index ('Coefficient Générique') for the flora as a whole. This is defined by the formula:

$$\text{Richness Index} = \frac{\text{number of genera} \times 100}{\text{number of species}}$$

To compare the flora with that of other areas the Sørensen Coefficient (K) is often employed (see e.g. Sheard, 1962):

$$K = \frac{200 \times c}{a + b}$$

where a is the number of species in one region; b, the number of species in the other; and c, the number of species which occur in both areas.

The list of taxa itself should be arranged so as to enable its users to extract information they require as easily as possible. The species can be arranged either according to an accepted taxonomic system (a flora is *not* the place to propose a new system) or alphabetically by genera. If a system which is not alphabetical is adopted, however, an index to the genera is essential. It is not usually possible to include full lists of synonyms for each species because of limitations of space but names used in previously published studies on the area should be included and cross referenced. Where imperfect states are treated only under the names of the perfect states the imperfect state names need to be cross-referenced to their perfect state name entries.

Citation of the place of publication of names is not usual in local floras but authors should ensure that the names they employ are the currently accepted ones and that the authorities given are correct. It is advisable to let a taxonomist check the nomenclature of a flora before the manuscript is finalized for publication. A flora is not the place to describe new species or to make large numbers of new combinations, but if for any reason either of these is necessary particular attention should be drawn to them in an abstract, summary or list of new names.

For each species data on its host, substrate or habitat, frequency and distribution within the area are desirable. Notes on variability and any other observations (e.g. seasonal occurrence, evidence of introduction or extinction, chemical components) can also be included with the treatment of each species. For rarer species full details of all collections are essential and should include the precise locality, date of collection, collector, collection number and herbarium where the material has been deposited (see also pp. 68–70). Details of representative collections of even the commonest species should also be cited. Records based on published reports not supported by specimens checked by the author or which may for some other reason be suspect should be clearly indicated as such (see p. 68).

If some system of units (e.g. 10 km squares) for recording has been employed the distribution of commoner species can be given by reference to these. The published entry for the species card illustrated in Fig. 13 was presented as follows (Hawksworth, 1969a):

> *Acarospora fuscata* (Nyl.) Arnold
> Millstone grit. Frequent.
> **26**: Stanton Park, 18––,? *Bloxam* MANCH; Youlgreave, 18––, ? *Bohler*
> MANCH; Stanton Moor, 20 vi 1965, *Wallace* BM. **35**: Nether Heage, 1966,
> *Hawksworth* DLH (399). **37**: Hare Edge, 14 iv 1967, *Hawksworth* DLH (697).
> Recorded from **98, 06, 08, 14, 16, 18, 21–28, 32–35, 37, 46.**

Floras including the above types of information are perhaps of most interest to the specialist and are not primarily designed to help students with the identification of taxa from the area. Floras for identification are necessarily lengthy and should also include keys to the orders, families, genera and species, and brief diagnoses of each. Line drawings and(or) photographs of representative species are also valuable in such works. Descriptions and illustrations included in a flora of this type must be based on material from the area under investigation and not adapted from monographs or based on specimens from other areas. If this is not done confusion might result in the future.

In mycological floras including plant pathogenic fungi a host-index is a valuable addition. The style of the host-index can be either alphabetical or follow an accepted taxonomic system (e.g. by host-families). The latter type is of more value as it brings together related host genera and species which may have pathogens in common. For examples of host-indices see the annual indices to the *Review of Plant Pathology* and the ten-year indices of the *Index of Fungi*.

If the flora is not arranged alphabetically a comprehensive index to at least the genera is essential. Any help received from specialists can be mentioned in an 'Acknowledgements' section and the groups with which they have assisted indicated. If a specialist only dealt with very few specimens this can be indicated in the text in parentheses after the relevant records.

The preparation of a flora is both time consuming and tedious and in order that the mass of data can be made readily available the final format is very important. This will often be dictated by editors or publishers but those who have freedom of choice are recommended to study the layout of as many as possible of the floras listed in Ainsworth (1971) before proceeding. Modern vascular plant floras are also a valuable source of ideas.

For further general information on the preparation of floras see Heywood (1958), Davis and Heywood (1963), Turrill (1964) and Wanstall (1963).

MAPS

MAPS are valuable additions to accounts of species as they show at a glance their known distribution. They form a particularly important part of monographs and revisions and are being increasingly employed in floras. Maps are particularly useful in the case of commoner species where there are too many records to list each one individually, and where allied species exhibit characteristic distributions. There is, however, little point in producing maps for cosmopolitan species or where so few records are available that the distributional patterns are not clear. For a general discussion of the preparation of maps for biological data see the Royal Geographical Society (1954). Many practical hints are included in the work of Hodgkiss (1970).

Collection of the data

For maps to be produced as accurately as possible precise details of the locality are essential (see pp. 29–30). Each record to be mapped can be entered on an individual record card (see Fig. 19) and filed until enough data have been accumulated for a meaningful map to be produced. If the cards are filed according to a grid system, countries or smaller geographical units (e.g. counties) this will facilitate plotting at a later stage.

It is essential that all records which are going to be plotted are correct and to forestall any incorrect identifications being entered, maps should be based only on specimens studied by the author or reports by reliable workers. Where it is not possible to confirm unlikely published records they should be omitted from the map and the reports mentioned in an accompanying discussion.

Records can never be absolutely exhaustive without a search of every herbarium and piece of literature and the compiler of a map must, like the flora writer, decide when to stop his compilation and finalize his map. The best guide to a map approaching 'completion' is when additional records obtained almost entirely fall within the already established range and add relatively little to the general pattern already established. If a worker had records only from Europe and then suddenly found a further twenty or so from South America he would clearly not yet be in a position to produce a reliable map.

Base maps and plotting

The base maps on which to plot distributions should be standard throughout a particular paper and not cover a larger area than is necessary. If a group of species confined to south-east Asia is being dealt with there is little point in using a base map of the whole world. Maps in Mercator's projection are most favoured by botanists but polar views are often valuable when dealing with circum-polar species (see e.g. Ahti, 1961). The Goode Base Map Series (edited by H. M. Leppard and produced by the Department of Geography, University of Chicago) is frequently used and gives a latitude and longitude grid, shows country boundaries, and indicates major cities and rivers (all without names). Base maps should always be as simple as possible and should omit all geographical names, contours and mountain ranges. Some grid system is, however, essential for the accurate plotting of data.

Distributions may be indicated in four main ways: (1) by blocking in or shading countries or counties from which there are records; (2) by placing dots over the precise locality; (3) by placing dots in a grid system; or (4) by a thick boundary line including all the localities.

For most purposes dot methods are most satisfactory as these give the most precise indication of the data available and some information as to the frequency of the taxon within different areas. Where species are very common a boundary-line method combined with dots in areas where the species is less common is often valuable. Insets drawn to a larger scale are helpful in this latter situation if the boundary-line method seems inappropriate. Grid systems are becoming increasingly popular and while they are somewhat less precise they avoid the problem of overlapping dots. In such cases the grid used should be fairly small, 10 km square being particularly popular in Britain and 50 or 100 km square units when the whole of Europe is considered. In grid systems one spot indicates one *or more* records from the particular square.

Examples of different types of distribution maps are shown in Fig. 14–16. The relationship between dots and boundary-line methods and the inaccuracy of the latter method are indicated in Fig. 17. Stearn (1951) gives a more detailed comparison of types of distribution maps and their relative value.

If areas of species overlap very slightly or not at all it is often possible to include two or more taxa on a single map by using different symbols (e.g. dots, squares, triangles, asterisks) but the use of many symbols on a single map is often confusing. Great care must be taken in selecting the scale of the map and the size of the dots. The larger the scale of the map and the smaller the dots the more accurate the final result will be. The dots should not be so small that they are likely to be overlooked; any likely to be overlooked because they are away from the main geographical area of a taxon may be emphasized by an arrow.

Very great care must be taken in plotting localities, and reference to latitude and longitude or some other grid system is essential during plotting. The tracing of locality names the author is unfamiliar with sometimes presents problems and ways of finding these are discussed below (see pp. 86–89). Any localities which are not precisely located should be omitted. If there is a record localized only to a country, province or county a spot should not be placed in the centre of that area and it is better to refer to such reports separately in any discussion accompanying the published map.

More data than the simple presence or absence of a record can be included in a map. Solid dots and open circles are often adopted to indicate periods

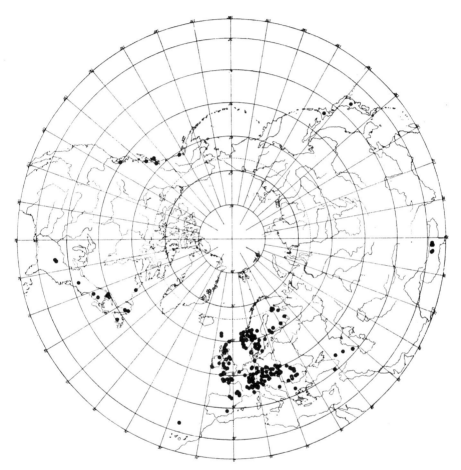

Fig. 14. Distribution map of *Alectoria bicolor* in the Northern Hemisphere in which all records are indicated by black spots (after Hawksworth, 1972a).

within which collections were made (see Fig. 16) or other differences (e.g. in chemical constituents). Some authors working with flowering plants attach symbols to their spots to indicate morphological variations (see Davis and Heywood, 1963) but this practice has so far not been widely adopted by mycologists.

In some instances it may be helpful to include other data on the map such as forest types, host ranges, geological formations, relative humidity contours, rainfall contours, or soil types, but these may serve only to confuse a map and

84

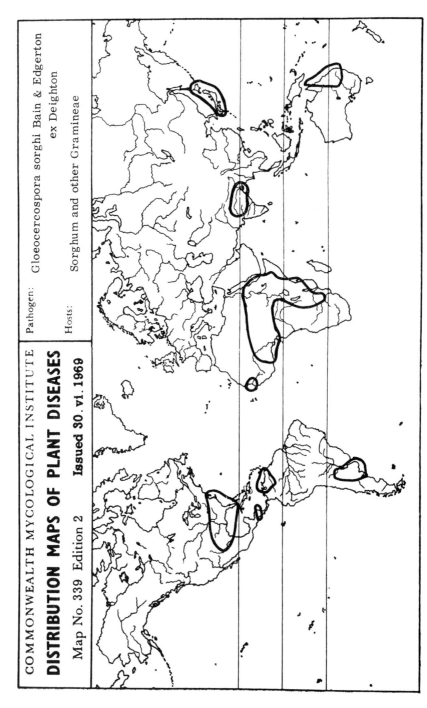

COMMONWEALTH MYCOLOGICAL INSTITUTE

DISTRIBUTION MAPS OF PLANT DISEASES

Map No. 339 Edition 2   **Issued 30.vi.1969**

Pathogen: Gloeocercospora sorghi Bain & Edgerton
             ex Deighton

Hosts: Sorghum and other Gramineae

Fig. 15. World distribution of *Gloeocercospora sorghi* indicated by encircling those countries (or states) from which it has been recorded.

it is often preferable either to publish separate maps with these types of data or to use overlays (see e.g. Perring and Walters, 1962; Cadbury *et al.*, 1971).

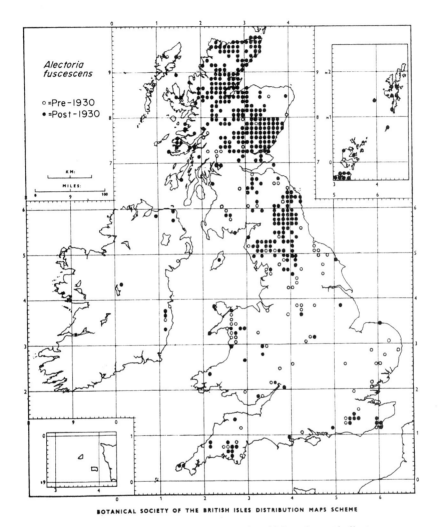

Fig. 16. British distribution of *Alectoria fuscescens* in which each spot indicates one or more records from the 10 km square unit of the national grid on which it is placed (after Hawksworth, 1972*a*).

Mapping Schemes

Following the success of the Botanical Society of the British Isles' Distribution Maps Scheme other scientific societies in Britain have initiated similar schemes (e.g. the British Lichen Society). These utilize standard record cards (Fig. 18) listing species with code numbers for each 10 km square unit of the British Isles (i.e. about 3,600 squares) which are completed by members of the society and forwarded to a central agency for sorting and checking. Maps may be produced from these and also 40-column (see Fig. 19) or 80-column individual record cards using a tabulator or computer adapted to print dots on maps. See Perring (1971) and Heath and Scott (1972) for further information on this method; a map produced by it is shown in Fig. 16.

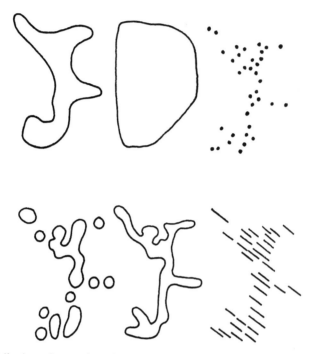

Fig. 17. Distribution of a species of lichen expressed by dots, hatching and outlines, all based on the same data (after Stearn, 1951).

A scheme for mapping macromycetes in Europe has been in operation for some time but no maps from it have been published yet. The British Mycological Society is in the process of establishing a mapping scheme similar to that of the British Lichen Society (see Rayner, 1973).

Tracing localities

As mentioned above, great care should always be taken in tracing localities so that they can be plotted as accurately as possible. Authors often have great difficulty in finding just where particular sites are when dealing with maps on a world scale. To trace localities gazetteers are necessary. At the moment there appears to be no comprehensive list of published gazetteers available. *The Times Index-Gazetteer of the World* (London, Times Publishing Co. Ltd., 1965) is the most useful single work for general use. The series of gazetteers published by the United States Board on Geographical Names are even more comprehensive and, like *The Times Index-Gazetteer*, give the latitude and longitude co-ordinates

A

	LOCALITY	Name P.W. James +D.L. Hawksworth

Grid Ref. 5 5 4 9 0 2

LOCALITY: Berry Head, nr. Brixham.

Date: Aug. 1972 V.C. No. 3

HABITAT: limestone cliffs, short turf, walls, palings and scrub.

V.C. S. Devon

Alt. 0-57 m Code No. —

0024	Acar atra	0326	*nobu*	0611	grif	0824	bach	1020	prem	1376	fuli
0034	cerv	0328	scop	0617	*tent*	0835	*crtsp*	1022	*tott*	1381	furv
0040	fuscata	0331	spha	0619	ligh	0838	*aniot*	1026 Lecania		1400	fuscoat f
0043	glau	0347	umbr	0632	pras	0841	fasc		aipo	1401	gris
0046	hepp	0366 Baeo plac		0634	pulv	0843	flac	1033	cyrtella	1403	gela
0048	macr	0368	rosc	0638	spha	0846	fragil	1038	*soya*	1405	glau
0072	sino	0370	rufu	0647 Catin		0848	fragr	1040	fusc	1408	goni
0074	smar s	0373 Belo russ			gros	0850	furf	1043	nyla	1410	*gran*
0075	lesd	0375 Biat camp		0654 Cave hult		0854	limo	1056 Lecanor		1417	hydr
0083	vero	0385	mori	0656 Cetr chlo		0856	mult		acto	1421	illi
0087 Alec bico		0388	ochr	0657	comm	0857	nigr	1072	atra	1430	jura
0090	chal	0403 Botr vulg		0659	eric	0860	poly	1078	badi	1432	koch
0091	fusc	0407 Buel aeth	1	0660	glau	0863	subf	1082	caes	1435	lapi
0093	impl	0409	*albo*	0663	hepa	0865	subn	1084	*cele*	1437	leuc
0095	lane	0413	atra	0666	isla	0867	*tonci*	1088	*camp*	1440	*limi f*
0096	nigr	0422	*cone*	0676	niva	0868	ceza	1090	carp	1441	sora
0098	ochr	0424	chlo	0678	norv	0869	*subg*	1094	chlaron	1442	limo
0100	pube	0427	disc	0680	pina	0871	tuni	1095	*chlarot*	1443	lith
0103	sarm	0452	*pane*	0682	sepi	0876 Coni furf		1097	cinerea	1444	lopa
0106	subc	0457	scha	0684 Chae aeru		0882Cono homa		1105	*confue*	1445	luci
0110 Anap cili		0460	*atot*	0685	brun	0883 Cori viri		1108	coni	1447	luri
0112	fusc	0462	subd	0690	chry	0885 Corn acul		1110	acul	1450	macr
0115	leuc	0465	verr	0691	ferr	0887	muri	1112	*cron*	1457	ochr
0117	obsc	0473 Cali abie		0710 Clad arbu		0888	norm	1122	*disp* 2	1459	oros
0127 Arthon		0480	lent	0712	baci	0894 Cyph inqu		1124	epan	1460	pant
	cinn	0485	sali	0714	bell	0898	sess	1128	*axpe*	1463	pelo
0129	didy	0488	viri	0716	botr	0900 Cyst nige		1134	fugi	1464	perc
0131	disp	0495 *Cate*		0718	caes	0904 Dermati		1138	gang	1466	phae
0138	impo		*aurantia*	0721	cari		quer	1140	gibb	1469	plan
0140	lapi	0496 aurantiaca		0722	carn	0908 Dermato		1142	heli	1470	poli
0145	luri	0498	caes	0725	cerv		cine	1152	intr i	1472	pycn
0152	punctif	0499	cerina	0727	*chlo*	0909	fluv	1153	sora	1473	quer

– – – – – FOLD – HERE – – – – – – –

0156	radi	0501	cerinel	0728	cocc	0910	*hopo*	1155	intu	1479	acab
0158	spad	0503	*chut*	0730	*conic*	0915	mini	1157	jame	1480	acal
0172 Arthop		0506	cirr	0732	conis	0918	rufe	1160	lacu	1484	semi
	alba	0508	*atb*	0736	corn	0919 Dime dilu		1162	laevat	1485	sila
0174	ante	0515	deci	0738	cris	0920	lute	1165	lepr	1487	sore
0178	bifo	0518	ferr	0750	digi	0922 Dipl caes		1174	mori	1489	spei
0188	cine	0520	gran	0754	*funo*	0924	gyps	1176	mura	1492	*stig*
0190	*conc*	0524	*hopp*	0758	floe	0926	scru	1178	pall	1498	subin
0195	fall	0527	*holo*	0760	*foli*	0943 Ente cras		1182	pini	1504	sulp
0199	halo	0532	lact	0762	*furc f*	0947	hutc	1184	polio	1506	sylv
0205	punc	0534	litt	0763	subr	0950 Ephe lana		1188	poly	1508	*symm*
0206	pyre	0536	lute	0764	glau	0956 Ever prun		1194	prev	1511	tayl
0209	salw	0538	*maci*	0765	gone	0959 Fulg fulg		1202	rupi	1513	temp
0211	saxi	0541	micr	0766	grac	0960 Gomp caly		1206	samb	1515	tenebrica
0237 Arthoth		0543	*muco*	0768	impe	0964 Gong sabu		1212	subcar	1517tenebricos
	ilic	0545	obli	0770	incr	0966 Graphina		1214	subcir	1519	tenebros
0242	ruan	0547	*coh*	0772	lute		angu	1217	subf	1525	tumi
0245	spec	0557	stil	0774	maci	0970	ruiz	1222	tene	1529	turg
0253 Baci arce		0558	teic	0780	ochr	0972 Graphis		1226	vari	1533	ulig
0257	beck	0559	*bot*	0782	papi		eleg	1231	verr	1541	vern
0265	chlo	0561	thal	0784	para	0975	scri	1239 Lecid		1552	wall
0269	citr c	0563	*vesi*	0786	pity	0984 Gyal flot			agla	1557	wats
0270	alpi	0570Candelaria		0788	*pool*	0985	fove	1241	alboc	1634 Lemp botr
0273	cupr		conc	0789	poly	0988	geoi	1263	assi	1652 Lepr cand
0277	*endo*	0572 *Candelariel*		0791	pyxi	0990	jene	1269	auri	1653	chlo
0283	frie			0793	rangifer	0994	trun	1279	caes	1655	*arou*
0291	herba	0574	cora	0794	*rangifor*	0996	ulmi	1285	cine	1656	*inco*
0296	inco	0577	*medi*	0796	scab	0998 Haem cocc		1287	cinn	1661	memb
0300	inun	0579	*uito*	0798	squa s	1000	elat	1291	coar	1665	negl
0304	lepr	0581	xant	0799	allo	1002	vent	1303	crus	1672	xant
0307	lign	0586 Catil		0800	stre	1004 Icma eric		1354	cyat	1676 Leptog
0310	mela		atro	0802	subc	1006 Iona epul		1355	deci		breb
0314	*cinov*	0588	bifo	0805	subu	1010	suav	1356	demi	1678	burg
0316	naeg	0590	bout	0809	tenu	1012 Lecanac		1358	dick	1633	cyan
0318	nits	0592	chal	0814	unci		abie	1362	didu	1692	lich
0322	phac	0594	chloros	0816	vert	1015	dill	1370	crra	1695	micr
0324	rube	0603	deni	0822	*Coll ausi*	1018	ploc	1373	frie	1703	pusi

Fig. 18. Record card used in the British Lichen Society's Distribution Maps Scheme completed for a locality in Devonshire; A, front; B, (see overleaf) reverse.

B

Grid Ref.							LOCALITY

2 0 9 4 . 5 6 .   Berry Head, nr. Brixham.

1708 satu	1903 pity	2032 spur	2200 cupu	2392 poly	2595 papu
1710 schr	1905 rubi	2034 veno	2202 demi	2403 umbi	2598 pyre
1713 sinu	1907 samp	2037 Pert	2204 derm	2405 viri	2610 Theloc laur
1718 tere	1910 Parmelia	albe a	2206 gela	2407 Rino atro	2624 Thelo prube
1720 trem	acet	2038 cora	2209 hens	2409 ~~bive~~	2626 Thelot
1732 Leptor epid	1912 alpi	2040 amara	2220 scot	2419 exig	lepa
1736 Lich conf	1916 aspe	2046 chiod	2226 thel	2425 luri	2633 Thro epig
1738 pygm	1920 cape	2048 cocc	2228 tris	2427 oxyd	2643 Toma gela
1744 Lith tess	1924 cetr	2051 cora	2236 Polyc musc	2430 robo	2649 ~~Toni erom~~
1750 Loba ampl	1926 cons	2053 dact	2237 Pori ahle	2433 soph	2654 cara
1752 laet	1928 crin	2055 deal	2238 ~~ahle o~~	2435 ~~subon~~	2660 coer
1756 pulm	1930 deli	2059 flavic	2239 carp	2439 teic	2664 lobu
1758 scro	1935 eleg	2060 flavida	2240 ~~pass~~	2446 Rocc fuci	2684 Umbi cyli
1764 Lopa pezi	1936 endo	2064 hemi	2243 guen g	2447 phyc	2687 deus
1766 Mass carn	1937 exas	2066 hyme	2244 curn	2448 Sarcog	2692 hype
1782 Mene tere	1938 furf	2068 lact	2246 luce	clav	2694 polyp
1790 Microg	1940 glab g	2070 leio	2253 lect	2449 priv	2696 polyr
musc	1941 fuli	2076 micr	2255 lept	2450 ~~regu~~	2698 prob
1798 Mycob	1946 incu	2078 mono	2260 oliv	2451 simp	2700 pust
sang	1950 isid	2080 mult	2265 ~~Prot immo~~	2453 Scle circ	2704 torr
1808 Neph laev	1952 laci	2098 pert	2266 incr	2454 Schi abie	2708 Usne arti
1810 pari	1954 laev	2104 pseu	2267 metz	2455 deco	2710 cera
1813 Norm pulc	1955 loxo	2106 pust	2268 ~~mont~~	2457 ~~Sole cand~~	2713 exte
1814 Ocel subt	1956 moug	2116 Petr clau	2269 ~~rupe~~	2459 holo	2714 iili
1815 Ochr andr	1958 omph	2118 Phae dend	2273 Pseu croc	2461 vult	2715 flam
1817 frig	1960 perl	2122 lyel	2275 thou	2464 Solo croc	2716 flor
1821 inve	1962 phys	2124 rami	2276 Psorom	2467 sacc	2718 frag
1824 ~~pers~~	1964 prol	2126 Phly agel	hypn	2468 spon	2719 fulv
1829 tart	1967 redd	2128 arge	2280 Psorot	2470 Spha frag	2721 glabres
1831 turn	1969 reti	2129 Physcia	scha	2472 glob	2722 hirt
1835 yasu	1970 revo	adgl	2309 Pyrenu	2474 mela	2726 rubi
1840 ~~Opeg atra~~	1972 saxa	2130 ~~adse~~	nitida n	2488 ~~Squa crus~~	2730 subf
1845 ~~cele~~	1974 sinu	2132 ~~eipe~~	2310 nitidel	2490 lent	2731 subp
1847 cesa	1975 soredian	2135 ~~sess~~	2312 Raco rupe	2494 Stau caes	2735 Verr aeth
1849 ~~ahen~~	1980 subaur	2138 clem	2316 Rama cali	2498 fiss	2742 aqua
1851 ~~conf~~	1981 subr	2144 gris	2321 curn	2500 hyme	2747 coer
1857 gyro	1982 subst	2147 ~~lept~~	2324 ever	2506 ~~supi~~	2755 ~~dufe~~
1859 herb	1984 sulc	2151 nigr	2328 fari	2508 succ	2757 elac

– – – – – – – – – – – – FOLD – – HERE – – – – – – –

1864 lith	1985 tayl	2154 ~~orbi~~	2330 fast	2514 Sten pull	2762 ~~glau~~
1866 lync	1986 tili	2160 pulv	2333 frax f	2516 sept	2765 ~~hoph~~
1868 moug	1988 tubu	2164 stel	2334 cali	2524 Ster dact	2767 hydr
1870 nive	1992 Parmeliel	2166 ~~tene~~	2338 obtu	2527 deli	2774 marg
1874 para	atla	2168 tere	2340 poll	2531 evolutum	2776 maur
1876 ~~pass~~	1996 cora	2170 tribacia	2343 sili	2538 micr	2780 micr
1878 pros	2004 plum	2171 tribacio	2345 subf	2542 pile	2784 muco
1885 rufe	2009 Parmeliop	2172 wain	2349 Rhiz alpi	2543 saxa	2786 ~~mura~~
1887 saxa	aleu	2174 Pilo dist	2357 cons	2546 vesu	2793 ~~nigr~~
1888 saxi	2010 ambi	2176 Placid	2359 dist	2550 Stic cana	2799 prom
1889 sore	2012 hype	cust	2363 gemi	2551 dufo	2812 ~~ophit~~
1890 vari	2014 Pelt	2178 Placop	2365 geog	2552 fuli	2815 stri
1891 verm	apht a	geli	2369 hoch	2554 limb	2819 ~~viri~~
1892 viri	2015 vari	2182 Placy	2372 lava	2556 sylv	2843 ~~Xant aure~~
1893 vulg	2017 ~~cani~~	flab	2374 leca	2559 Syna symp	2845 cand
1895 zona	2018 coll	2186 ~~nigr~~	2376 lind	2564 Telo flav	2847 eleg
1896 Pach corn	2019 hori	2191 tant	2379 obsc o	2569 Tham verm	2849 ~~pari~~
1899 Pann micr	2023 ~~poly~~	2193 trem	2380 redu	2573 Thel deci	2851 poly
1901 nebu	2024 prae	2197 Polyb	2382 oede	2583 inca	2854 Xylo abie
1902 pezi	2027 rufe	albi	2388 petr	2591 micr	2864 viti

Additions:-

1 = as B. epipolea.

2 = with f. *albescens* and f. *verrucosa*.

3 = with var. calva.

Toninia cervina Lönnr.

British Lichen Society 1968

Fig. 18—*cont.*

which are the best way of plotting localities on maps. Many more local gazetteers are available such as the *Ordnance Survey Gazetteer of Great Britain* (Chessington, Surrey, 1953) and Bartholomew's *Gazetteer of the British Isles* (Edinburgh, J. Bartholomew & Son Ltd., 1966) for the British Isles; *The Imperial Gazetteer*

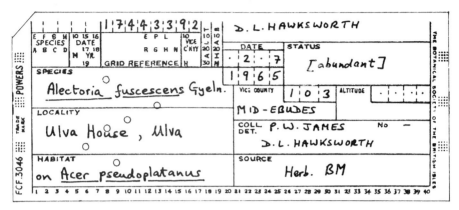

Fig. 19. One of the punched individual species 40-column record cards used in the production of the map reproduced here as Fig. 16.

of India (Oxford, Clarendon Press, 26 vols., 1907–09); the *East Africa-Index Gazetteer* (Cairo, Survey Directorate, 3 vols., 1946); the *Gazetteer of Canada* (Canadian Permanent Committee on Geographical Names, many volumes undergoing revision); Merriam *et al.* (1972); etc. The *Stanford Reference Catalogue* (London, Edward Stanford Ltd., 1969) lists currently available maps and gazetteers.

In tracing localities in gazetteers always bear in mind that the spelling on a packet may be incorrect or obsolete and try obvious likely alternative names and misspellings. This is a common failing of collectors working in areas they are not very familiar with when names may be heard by word of mouth, written down quickly in the field, and never checked against spellings accepted on maps.

SOCIOLOGY

SOCIOLOGY is the study of communities, phytosociology of plant communities in general, and mycosociology of fungal communities in particular. Just as individuals which are morphologically similar are grouped into species, groups of morphologically distinct species characteristic of particular habitats can be grouped into communities. Where these communities are sufficiently distinctive they can be given latinized names based on their characteristic species which are followed by author citations. Phytosociology is often studied by ecologists lacking specialist knowledge and consequently such studies often tend to overlook critical and minute species and include incorrect identifications. For such studies to be comprehensive and accurate a sound knowledge of taxonomy is prerequisite and taxonomists are therefore the ideal workers to conduct such surveys. A comprehensive introduction to most aspects of phytosociology is given by Shimwell (1971) and this work should be consulted for further information on aspects discussed in this section.

Mycosociology has received relatively little attention as compared with sociological studies carried out on vascular plant, bryophyte and lichen communities. This arises from two main problems, the number of fungal communities which often occur within a relatively small area and within single vascular plant communities, and the occurrence of fungi as mycelial states not producing perennial sporocarps. Any mycologist who has done field work soon appreciates the complexity of the situation, noting how particular fallen trees or fruits and particular hosts have their own characteristic species, and how the larger fungi producing sporocarps in a particular forest or pasture vary from season to season and year to year.

Most of the work on mycosociology has consequently been limited either to listing the species which occur on different hosts or in different habitats, or to ascertaining the habitat range of particular species. Some authors have carried out studies on the fungi occurring within phytosociological taxa based on the vascular plants (e.g. Eihellinger, 1973; Lisiewska, 1972; Majewski, 1971; Thoen, 1970; Ubrizsy, 1972); this is a particularly valuable approach in studying macromycete floras of, for example, woodlands. Höfler (1938) and Hueck (1953), however, consider that fungal communities can be studied and named independently of other communities because they are dependent on factors distinct from those such as illumination which control the communities within which they occur. Lichen and bryophyte communities on trees have been named independently of tree communities for many years. Relatively little progress has yet been made in the naming of fungal communities according to the principles widely adopted in other plant groups. Surveys of the literature are given by Cooke (1948), Hueck (1953) and Apinis (1972) and some further pertinent references are listed in Ainsworth (1971). A valuable summary of problems in mycosociological studies is provided by Barkman (1973).

The description and naming of lichen communities is, in contrast, in a relatively advanced stage, so much so that no detailed floristic account of an area can now be considered complete without a section on sociology. It is also useful to be able to indicate in monographic and revisionary studies in which communities particular species occur. The remainder of this section discusses the classification systems, rules of nomenclature and description and delimitation of communities with particular reference to those including or dominated by lichens.

Classification Systems

There are two principal systems of phytosociological classification, the Uppsala-system established by G. E. Du Rietz, and the Zürich-Montpellier-system established by J. Braun-Blanquet. Both are mainly based on similar criteria but the ranks are given different names. The basic unit of the Uppsala-system is the 'union' and as used is roughly equivalent to the 'association' of the Zürich-Montpellier-system; these are grouped into 'federations' and 'alliances', respectively. Unions and associations end in the suffix '-etum' (e.g. *Parmelietum furfuraceae* Hilitz.) and federations and alliances in the suffix '*-ion*' (e.g. *Physodion, Xanthorion*). More ranks are used in the Zürich-Montpellier system, some of lower values than associations (e.g. subassociation, variant) and others higher (e.g. order, ending in the suffix '*-etalia*', as in *Leprarietalia*).

These two systems also differ somewhat in their emphasis and usage of characteristic species as criteria for classification; the Uppsala-system stresses dominance while the Zürich-Montpellier-system emphasizes fidelity (i.e. species confined to particular communities but not necessarily constantly present). Shimwell (1971) provides a comprehensive comparison of these and other systems. The Uppsala-system, modified in various ways, has been used mainly by British, North American and Scandinavian authors, and the Zürich-Montpellier-system has been applied principally by authors from most other European countries. The Zürich-Montpellier-system is now being accepted more widely and is recommended as the most satisfactory for general use.

Nomenclature

Names of communities are based on the Latin names of one or more of their characteristic, but not necessarily dominant, species. Words such as '*typicum*' and '*atlanticum*' have frequently been applied at lower ranks but are not now recommended.

Just as a complex set of internationally agreed rules is accepted for the naming of genera and species to stabilize and control their nomenclature, rules have been proposed* for the naming of vegetation units. These rules are based on the assumption that names of communities should only be changed as a result of revisionary work, and not because they appear atypical in the light of increasing knowledge, or of nomenclatural changes in names of genera and species included in their names unless to do otherwise would lead to error or confusion. Consequently the name of a community may be based on a species which is not present throughout its geographical range or constantly present in it, or on a nomenclaturally obsolete name. The association *Lecanoretum pityreae* Barkm., for example, is still so named although *Lecanora conizaeoides* Nyl. ex Cromb. is now known to be the correct name for the species formally called *L. pityrea* Erichs.

A detailed set of 96 rules was proposed by E. Meijer Drees in 1951, an abridged English version of which appeared in 1953 (Meijer Drees, 1953) and these are now generally followed with a few minor modifications (Barkman, 1953, 1958). The rules themselves largely parallel Articles in the International Code of Botanical Nomenclature and only a few of the more important are referred to here. The date proposed as the starting point for the names of vascular plant communities is 1 January 1900 and Barkman (1958) proposed Allorge (1922) as that for epiphytic communities. Names are considered as valid if given only one name based on that of one or more constituent species, provided with an indication of rank, accompanied by at least one complete record (i.e. a list of species in the community with presence degrees), and if they are not later

---

* No internationally authorized rules for the naming of plant communities have yet been published but some are in the course of preparation (Barkman, 1973, *in litt.*).

homonyms of previously validly published community names. One difference from the International Code of Botanical Nomenclature is that names in all languages are accepted and latinized without change in the author citation (e.g. the 'Association à *Parmeliopsis ambigua*' Hilitz. is corrected to the association '*Parmeliopsidetum ambiguae*' Hilitz.).

Where the names of communities are changed in rank the name of the describing author is placed in parentheses and that of the transferring author appended (e.g. the association *Leprarietum candelaris* (Mattick) Barkm. based on the '*Lepraria candelaris*-Gesellschaft' Mattick; the *Dichaeneetum fagineae* (Barkm.) D. Hawksw. based on the *Dichaena faginea*-sociation Barkm.). It should be noted that in the Zürich-Montpellier-system 'subassociation' and 'variant' have the same rank, including non-geographically and geographically based subdivisions of associations, respectively.

The type method is fundamental to the International Code of Botanical Nomenclature and this is also used in the naming of vegetation units. Each higher unit is typified by a lower, the lowest being typified by a type record. This point was not emphasized by Meijer Drees but is considered essential by Barkman (1953) to avoid confusion when units are divided or remodelled. Type records provide a clearer indication of what the author understood his community to be where the stands in his original table show considerable variation. Descriptions of new communities are usually presented as a table in which each vertical column represents a particular stand; one of these vertical columns should be designated as the 'type-record' of the community name (see Fig. 20).

Description and delimitation

In commencing a sociological study of an area the first step is to become familiar with all the species present in it so that they can be readily recognized in the field. It will soon become apparent that different groups of species characterize different habitats and then it is time to collect as much data as possible. To do this quadrats (and/or rectangles in the case of epiphytes) are placed over representative stands of the community. The size of the quadrat depends on the homogeneity of the community and whilst one of 10×10 cm may be adequate for epiphytes, 10×10 m or even larger units may be required for studying macromycetes on woodland floors. Kershaw (1964) and Shimwell (1971) give accounts of the effect of quadrat size on the resultant data. In investigations of periodically appearing fungi 'permanent quadrats' marked by stakes in the ground are essential and should be examined at as many seasons over as many years as practical. Six to eight visits each year for three years has been suggested as minimal for larger fungi in woodland soils, and Barkman (1973) recommends a period of 5–10 years as a result of his studies on the macrofungi associated with juniper scrub in the Netherlands.

The Zürich-Montpellier school employ a 'minimal area' concept (i.e. the least area which includes all the characteristic species) to establish the optimum quadrat size for a particular community when dealing with flowering plant communities but this concept seems to have relatively little application when dealing with epiphytic stands.

It is convenient to record the data from each quadrat on a loose-leaf sheet and place the details of the locality together with further data (e.g. rock-type for saxicolous communities and tree species for epiphytic communities, aspect, angle of inclination, pH of substrate, percentage uncolonized substrate, date, name of recorder). The species present should then be determined and listed in

Table XXXIII Graphinetum platycarpae

	1	2	3	4	5	6	7	8	Average:
No. of vegetation record	1	2	3	4	5	6	7	8	
Fieldno. of record	1822	1823	1930	1831	1707	1695	1721	1709	
Tree species	A	A	I	A	Fr	F	F	Fr	
Height in dm.	13.18	6.18	5.15	10.20	7.18	5.12	7.18	10.18	
Direction of exposure	SW.NW	E.NE	SW	all		N	E.SE	S.SE	
Inclination	77.80	76.89	65.79	83.95	88.96	70.79	89.94	98	
Sample plot surface (sq. dm.)	10	25	15	40	55	22	33	34	
Number of species	11	16	12	14	7	10	14	11	11.9
Total cover % crustaceous layer	60%	90%	100%	90%	100%	60%	100%	100%	88%
Faithful species:									
Graphina platycarpa	.	.	2.3 f	1.1 f	2.4 f	+.3 f	4.3 f	.	IV 190
Graphis elegans	8.3 f	1.2 f	.	2.7.3 f	+.2 f	.	.	.	III 145
Phaeographis dendritica	(+.2 f)	2.2 f	+.2 f	4.5.4 f	.	.	.	.	III 230
Differential species of the ass.:									
Pertusaria amara [1]	(+.2)	+.2	2.3	2.2	.	+.2	.	.	IV 139
Trentepohlia abietina	1.2	2.2	+.2	+.1	.	.	.	.	III 95
Pertusaria pustulata	+.2 f	2.1 f	+.2 f	.	.	.	.	.	II 45
Didymosphaeria micula	.	.	2.4 f	1.2 f	.	.	.	.	II 103
Lecanora subrugosa	.	.	.	.	+.2 f	2.3 f	2.2.4 f	2.3 f	III 115
Trentepohlia umbrina	.	.	.	.	2.4	2.3	4.5	2.4	III 180
Pyrenula chlorospila	.	.	2.2 f	.	.	.	.	+.3 f	II 40
Hysterium pulicare	4.1 f	.	.	+.1 f	.	.	.	.	II 3
Opegrapha pulicaris	.	.	2.2 f	+.1 f	.	.	.	.	II 40
Faithful species of the alliance.									
Graphis scripta	.	2.3 f	1.2 f	.	.	2.2.3 f	2.3.4 f	2.3 f	IV 220
Pertusaria leioplaca	(+.2 f)	1.2 f	.	1.1	.	+.3	.	+.2	III 30
„ pertusa	.	.	+.1	.	+.3	+.3	.	+.3	III 10
Dichaena faginea	.	.	.	.	.	1.1	2.4	.	II 105
Companions:									
Phlyctis argena	.	.	.	2.3	.	2.3	+.2	2.3	III 115
Lecanora expallens	.	.	.	.	+.2	+.2	2.2.3	2.2	III 80
Frullania dilatata	.	1.1↓	.	2.2↓	.	.	.	+.2↓	II 50
Arthonia radiata	1.2 f	.	.	.	.	+.2 f	.	.	II 15
Lecanora subfuscata	2.2 f	+.1 f	.	.	.	.	.	.	II 40
Lepraria candelaris	.	+.1	.	+.1	.	.	.	.	II 5
Parmelia caperata	.	+.1↓	.	+.1↓	.	.	.	.	II 5
„ trichotera	.	+.1↓	.	2.1↓	.	.	.	.	II 3
Arthonia cinnabarina	+.2 f	4.2 f	.	.	.	.	.	.	II 8

Fig. 20. Original table accompanying the description of the association *Graphinetum platycarpae* Barkm. (after Barkman, 1958); record 4 is the type of the association and record 6 is the type of the variant *lecanorosum expallentis* Barkm., cover and frequency are expressed according to the Braun-Blanquet system and fidelity is indicated by the Roman numerals in the right hand column.

tabular form and their abundance (cover, dominanz, dominance) and sociability ascertained. Vascular plant, bryophyte and algal associates should also be treated similarly where they form parts of the same community. Abundance can be scored according to one of three main systems:

(1) *Braun-Blanquet scales*. This system, used widely by the Zürich-Montpellier-school and modified for the fungi by Hueck (1953), is shown in Table 3.

Table 3. *Braun–Blanquet scales for estimating abundance and sociability as modified for the fungi by Hueck (1953)*

Abundance	Sociability
− no estimate	1 isolated
+ solitary or a few odd specimens	2 small groups or tufts
1 plentiful but sparse; small cover value	3 small patches or cushions
2 numerous; cover less than 25% of area	4 small colonies in extensive patches
3 fairly abundant; cover 25–50% of area	5 crowded in more or less pure stands
4 abundant; cover 50–75% of area	
5 very abundant; in masses, cover over 75% of area.	*Additional marks added to Sociability scores*
	o in rings
	− [under mark] in mixed patches
	1 only one such group in stand

In scoring abundance and sociability the numbers are separated by a full stop; '2.2' for example indicates a species which is numerous occurring in small groups or tufts, and '1.2.1' one which is sparse, occurring in small groups or tufts but with only one such group present (see Fig. 20).

(2) *Domin-scale* (see Ainsworth, 1971; Shimwell, 1971). This scale is an extension of the Braun-Blanquet scale but with 0–10 points. Scores in this scale, which is widely used, are readily interconvertible with the Braun-Blanquet scale (see Kershaw, 1964) and may be used together with sociability scores.

(3) *Percentage cover*. Cover may also be expressed as a percentage estimated either by eye or by careful measurements using, for example, point quadrats, tracings and graph paper, or gridded quadrats (noting the species present at line intersections) (see Shimwell, 1971). Because these latter methods are more time consuming they are used less frequently than (1) and (2).

It is essential that the utmost care is taken in identifications and any doubtful specimens should be removed for later study and placed in carefully numbered packets in the field.

When a large number of records have been obtained from all habitats they should be sorted into groups according to their similarity (i.e. numbers of principal species in common) so that the communities present can be determined. This must remain largely a matter for subjective judgement although numerical analyses (see p. 96) may prove valuable. Most will be found to intergrade to some extent (i.e. have some species in common) and be most appropriately interpreted as noda (i.e. peaks) on a continuum related to habitat vectors. Each nodum is usually characterized by a few predominant species which are absent or much rarer in other noda and found to be restricted to particular substrates (e.g. rock-type, hosts) and limited by one or more microclimatic factors (e.g. substrate pH, illumination, relative humidity).

All communities to be recognized should be as sharply delimited from each other as is possible so that stands not readily referable to one of two are much rarer than the communities themselves. In general terms it is better to recognize too few noda than too many relatively unclear ones; divergences and inter-gradations can always be discussed in text. To establish that a scheme being proposed is practical it is valuable to take colleagues to the area to see if they are able to pick the same communities out readily in the field. Lichen communities should all be recognizable quickly in this way.

Application of names

The naming of lichen communities, particularly in continental Europe, is now in a fairly advanced stage as compared with that in most other parts of the world, and so noda discovered in Europe will probably be found to have been already described and named. For accounts of the described communities in this area see Barkman (1958, 1962, 1966) for epiphytic, Klement (1955, 1960) for saxicolous, and Gams (1927) and McVean and Ratcliffe (1962) for terricolous communities, together with the references they cite. Recent work is listed in Culberson (1951→) and *Excerpta Botanica, Sect.* B (1951→); and *Vegetatio* (1948→) often contains papers of interest to lichenologists.

In matching noda with particular named communities great care is necessary. Original descriptions and tables should always be studied, paying particular attention to the substrate, species composition (including non-dominant but faithful species) and its known geographical distribution. Some communities occur in several habitats whilst others are restricted to particular ones. One of the characteristic species may not be present throughout a community's geographical range and consequently absent in one area where it may or may not be replaced by another (often related) species. In such instances it is clearly preferable to recognize the nodum at a lower rank (e.g. sociation, variant) than to describe it at a higher one (e.g. union, association).

Some lichen species in a community may be more susceptible to damage by air pollution and agricultural sprays and dressings than others and so atypical communities often occur in areas where such man-made factors are important (see Ferry *et al.*, 1973). Bark acidity may also be affected by air pollution causing species to occur on tree species other than those on which they are encountered in unpolluted areas. Consequently the application of community names in such areas is difficult and is often only possible by reference to those present on unaffected but otherwise identical substrates in adjoining relatively unpolluted areas. When working in polluted areas a great deal of caution is therefore necessary.

Relationships

It has already been mentioned that communities tend to intergrade with one another and this can be due either to variation in some habitat factor (e.g. illumination) or to succession (i.e. replacement of one community by another over a long period of time). Accurate studies on relationships require extensive measurements of pertinent ecological factors and consequently are usually outside the scope of floristic accounts. Some attempts to understand the relationships between communities is, however, desirable in sociological studies. Often these may be obtained by comparing recent and older substrates (e.g. saplings v. mature trees, recent v. ancient glacial erratics or moraines). Photography and tracings of fixed 'permanent quadrats' regularly over many years are the most satisfactory way of investigating successional changes. For examples of the application of the photographic method see Frey (1959) who studied changes in lichen communities on different substrates over periods of 18–19 and 33–34 years; and of the tracing technique see Brodo (1968) and Gilbert (1971). Lichens grow so slowly that long-term studies are necessary to obtain reliable data and so if an author wishes to investigate his communities in this way it is clear that such observations must be started as soon as possible after work on the flora has commenced.

In sociological accounts it is better to be cautious in interpreting relationships between communities so as to avoid introducing concepts which are based largely on speculation and which may well prove erroneous in the light of subsequent long-term studies in the same area. From careful field observations, however, it is often possible to suggest relationships apparently controlled by environmental variables such as illumination or substrate acidity but these are not to be confused with successional trends.

Numerical methods

Numerical (quantitative) methods, often requiring computer facilities, have been used extensively in the study of vascular plant communities (Grieg-Smith, 1964; Kershaw, 1964) and to a lesser extent in the macromycetes (e.g. Parker-Rhodes and Jackson, 1969) and lichens (e.g. Yarranton, 1967; Fletcher, 1973) and provide a means of grouping (clustering) individual records and assessing their affinity on a numerical basis. Many of the methods employed are similar to those used in cluster analysis techniques in numerical taxonomy (see pp. 102–104) and are consequently not discussed in further detail here.

Such studies usually require data from a very large number of stands and consequently are too time consuming to form parts of floristic accounts. They do, however, provide a field well suited to a research student able to work in conjunction with a taxonomist working on the area who is able to point to communities whose relationships seem complex and are suited to analysis by such methods.

Publication

Sociological studies fall naturally within the scope of regional floras (see p. 79), indeed an account of the flora of any area can hardly be considered comprehensive without some mention of the species and communities most frequently encountered or characteristic of particular habitats. In mycological floras this is often limited to lists of species occurring in different habitats (e.g. types of woodlands, grasslands) or host-indices. In the case of lichen floras, however, particularly in Europe, some attempt to ascertain the names of the communities present should be made and a study of some recent lichen floras (e.g. Laundon, 1967; Massé, 1966; Hawksworth, 1969, 1972b; Wirth, 1972) will indicate how this may be approached.

The mass of data collected in the course of phytosociological surveys cannot usually be published *in toto* in a paper because of high printing costs and authors will be required to keep the number of tables and lists they present to a minimum. Some lichen associations (e.g. *Cladonietum coniocraeae* Duvign., *Lecanoretum dispersae* Laund.) are extremely uniform in composition and so there is often no necessity to publish tables for these. Tables are, however, essential in the case of atypical communities discovered or ones being described for the first time. On each table sufficient numbers of records ('relevé's', 'Aufnahmen') must be included to provide a clear indication of the range of variation in the community and 6–10 records have been recommended as minimal. In the case of new communities being formally described the type record should be indicated either by underlining (*cf.* Fig. 20) or a footnote (e.g. Hawksworth, 1972b), and the species grouped according to the Braun–Blanquet concept of 'fidelity'; i.e. constancy (see Table 4) in and faithfulness to the community.

A detailed discussion of fidelity is outside the scope of this book, but further information is readily available through the works of Goodall (1952, 1953) and Shimwell (1971). The five classes of fidelity which are usually distinguished are summarized in Table 4. Fidelity values are usually indicated in the right-hand

Table 4. *Fidelity and constancy classes and their definitions*

Class	Fidelity	Constancy
V	Species ± restricted to the particular vegetation type	Constantly present (in 81–100 % of records)
IV	Species with a strong preference for the particular vegetation type but infrequently found in other types	Mostly present (in 61–80 % of records)
III	Species frequently occurring in other vegetation types but usually with their maximum frequency in one	Usually present (in 41–60 % of records)
II	Species lacking a preference for particular vegetation types	Seldom present (in 21–40 % of records)
I	Species which are rare or 'accidentally' occur in the particular vegetation type	Rarely present (in 1–20 % of records)

column of tables (*cf.* Fig. 20). More sophisticated methods of assessing fidelity are reviewed by Grieg-Smith (1964). The 'Sørensen coefficient' (K), often also used in comparing floras of different regions (see p. 179), can also be used to provide a numerical assessment of the affinities between communities (Grieg-Smith, 1964).

Latinized community names are either placed in italic type (e.g. *Opegraphetum fuscellae* Almb.) or in Roman type with spaces between each letter (e.g. Opegraphetum fuscellae Almb.).

When new names or transfers in rank are used 'ass. nov.', 'stat. nov.' etc. should be placed after the author citation where the new community is described or transfer made to render the nomenclature quite clear. Where new names have to be chosen they should conform to the requirements of Meijer Drees (1953) as far as possible, and be as short as the name(s) of the characterizing species permit.

Discussions of probable relationships between communities are often most appropriately presented as a separate section following the description of the communities present. In the preparation of phytosociological data for publication the general points discussed below (pp. 116–123) should also be considered.

SOME MORE COMPLEX TECHNIQUES

IN addition to studying fungi and lichens by the procedures outlined in the earlier sections of this book, some other more recent and more complex techniques are now available to the taxonomist. The systematist is concerned with any aspect of an organism which might lead to a better understanding of its characteristics and affinities. While directly observed morphological and anatomical criteria continue to provide the principal characters on which taxa are defined, data obtained from other approaches are often of considerable value in supplementing this information. Chemotaxonomical, numerical, serological and ultrastructural studies, for example, can lend support to the separation or union of taxa suggested by more orthodox methods, and indicate possible affinities which might not otherwise have come to mind.

This section reviews briefly some of the more complex techniques which have been employed in the fungi and lichens, noting their value and the types of information which may be derived from them. The employment of a particular technique in a family or genus for the first time frequently leads to data of interest to the taxonomist. It is a common failing of workers using some of these techniques, however, to overemphasize their importance. At the present stage of our knowledge none of the methods described here merits such stress, and data obtained from them should consequently be interpreted as material from which the taxonomist can derive additional data to consider in the formulation of taxa.

All the techniques summarized below are not, of course, applicable to every genus and group of species. The systematist embarking on any monographic or revisionary studies should, however, consider if any might be potentially useful to him and make some preliminary studies to ascertain if any are worth pursuing further. The more aspects of a group a systematist can consider the more comprehensive is his taxonomic treatment likely to be.

Chemotaxonomy

Chemical characters have been employed in lichen taxonomy since two papers of Nylander published in 1866. As pointed out on p. 55 chemical components of lichens form an essential part of all modern systematic studies on those groups which produce characteristic secondary products. Thallus reagent tests (p. 20) and microcrystal tests (p. 24) are extremely valuable aids in the routine identification of specimens but, as these tend to overlook some lichen substances, more sensitive techniques, particularly that of thin-layer chromatography (see p. 24) have to complement them in critical work and revisionary studies. Gas chromatography and mass spectrometry are also of value to the taxonomist but he may have to obtain the help of an organic chemist in interpreting his data. For further information on the chemical components of lichens and the methods for determining them see C. F. Culberson (1969, 1970, 1972), Culberson and Kristinsson (1970), Huneck (1968, 1971), Huneck and Linscheid (1968) and other references cited in Hawksworth (1971a).

By means of microcrystal tests and thin layer chromatography the chemical components of small fragments of herbarium specimens can be ascertained, so enabling the taxonomist to examine numerous collections of a taxon and work out the chemical variations within it. Various types of chemical variation occur in lichens and require different treatments. Some chemical differences are considered to justify the recognition of distinct species when related to geographical factors whilst others are of no taxonomic importance at all (i.e. chemotypes; see p. 45). Species rank, for example, is most appropriate for cases where one or more lichen products is replaced by one or more chemically very dissimilar compounds where there are major geographical correlations; that of variety for ones in which closely related compounds are involved where there are more local geographical and(or) ecological relationships (see p. 44); and no rank at all (chemotype) for instances where an unreplaced substance (i.e. an accessory) occurs sporadically through the range of a morphological species —i.e. chemical characters should be treated in the same way as variations in morphological and anatomical features. Where clear patterns of chemical variation are found the morphological characters of the chemical populations should be compared very carefully as in some cases correlations between chemical races and previously overlooked morphological variations have been discovered. For discussions of the taxonomic treatment of chemical races in lichens see W. L. Culberson (1969), Culberson and Culberson (1973), Hawksworth (1969a, 1970, 1973a, b), Hawksworth and Chapman (1971), Neelakantan and Rao (1967) and Poelt (1972). Culberson and Culberson (1970) discuss their phylogenetic significance, and Huneck (1971) reviews the available data on their biosynthesis.

In contrast to the lichens the non-lichenized fungi have only rarely been the subject of thorough chemotaxonomical investigations. The use of reagent tests in the Agaricales (see Watling, 1971) is in some ways comparable to the use of reagent tests in the lichens but while in the lichens the chemical structure of the compounds providing the reactions is usually precisely known, in the Agaricales it usually is not. For this reason reagent tests in this group need to be applied with caution as taxonomic criteria.

The lichen products which have been found to be of the most taxonomic interest are deposited in relatively large amounts on the outer surfaces of hyphae and can easily be extracted with acetone. The secondary metabolic products of most other fungal groups, however, are frequently within the hyphae themselves and this renders their isolation and identification more difficult. It is evident from the few chemotaxonomical studies that have been carried out in fungi other than lichens (see e.g. Olá'h, 1970; Arpin, 1969) that they can prove to be of considerable taxonomic interest. When dealing with fungi which produce characteristic colours, reactions (see Watling, 1971; Zoberi, 1972) or odours, steps should be taken to ascertain the chemical nature of the substance concerned. Where members of particular groups are already known to produce secondary metabolites which might prove of taxonomic interest, methods of determining these and establishing their occurrence in other taxa of the group should be looked into, and where feasible these data used as additional taxonomic characters. Valuable summaries of the metabolic products so far known in the fungi are provided by Shibata *et al.* (1964) and Turner (1971). Studies on the DNA base composition of fungi may eventually also yield information of interest to taxonomists but so far little progress has been made in this field (see Bertoldi *et al.*, 1973).

Cultures

Cultural studies have long been recognized as an important aspect of the study of microfungi (see pp. 33–35). There are, however, many species which have never been grown in artificial culture. The comparison of isolates from different hosts and(or) substrates grown on the same media under identical conditions of illumination and temperature provides a means of assessing the extent to which their morphological characters are affected by the host or substrate. In the case of perfect state fungi the germination of single ascospores or basidiospores frequently gives rise to conidial states which may or may not be already known. Imperfect states provide a useful aid to studies of perfect state fungi as differences in them can support separations and unions of taxa suggested by other characters. When working with perfect state fungi attempts to establish if any imperfect states occur should be made whenever possible. Even if the perfect state may not itself be produced, or produced only rarely, in artificial culture its imperfect state (if any occurs) may develop readily on an appropriate medium. Conversely, when working with imperfect state fungi, it is useful to try to grow them on a range of starvation media in addition to those on which they thrive and leave them for several months to see if a perfect state starts to develop.

When working with cultures comparisons of physiological parameters (growth rates on different media, temperature and pH optima etc.) are also sometimes of considerable interest and for this reason these data should be included in descriptions and accounts of species in culture (see p. 54).

The isolation of the fungal and algal components of lichens is now feasible (see p. 34) but the information derived from studies on isolated components has so far proved of rather little systematic value although developments in this field may be expected. In particular, rather little is known about the specificity of algal species within lichens. There are indications that in some lichen species different specimens contain different species of algae of the same algal genus. It is only by studying the algal components in pure culture that their taxonomy can be investigated. For examples of studies of algal components in connection with taxonomic studies in lichens see Degelius (1954) and Uyenco (1965), and for methods of isolation and culture see Ahmadjian (1967*a*, *b*).

Methods for growing intact lichen thalli in culture are also now being developed (see p. 34) and it is clear that these will eventually prove of considerable value in taxonomic investigations, enabling the rôle of environmental variables in the determination of thallus form to be investigated.

Cytogenetics

The genetical and cytological aspects of fungal taxonomy have been reviewed briefly by Lange (1968) and more detailed accounts of our present knowledge of the genetics of fungi are provided by Fincham and Day (1971) and Esser and Keunen (1965). As pointed out by Jinks and Croft (1971), however, whilst studies of the genetics of fungi have led to extremely important contributions in the understanding of genetic mechanisms, they have so far not been used extensively in the taxonomy of the fungi themselves. Cytological studies in the fungi have failed to assume the important rôle which such studies have come to play in some other groups of organisms, particularly the vascular plants (see Stebbins, 1950; Davis and Heywood, 1963).

By studying dividing nuclei in vegetative hyphae and in the process of meiosis in developing asci and basidia, by fixing them and employing chromatin specific stains (e.g. Giemsa after N hydrochloric acid hydrolysis; see Commonwealth Mycological Institute, 1968, Duncan and Galbraith, 1973), it has been possible to determine the chromosome numbers and nuclear status of some fungi (see e.g. Rao and Mukerji, 1970; Punithalingam, 1972). So far chromosome numbers have found little application in taxonomy in the fungi but as more species are studied they may be expected to prove of increasing value, particularly at the generic and family levels. In the case of the lichens, however, so few chromosome counts have been reported that it is not possible at present to assess any value they may have.

A further aspect of genetical studies which is of some taxonomic interest is the occurrence of homothallic and heterothallic species. When cultural studies on Ascomycotina are being carried out it is of value to determine if they are homo- or heterothallic. Where heterothallic taxa are being considered it is of value to try to mate opposite strains, when the production of ascocarps can prove their identity. In some cases mating experiments of this type can also confirm that previously described heterothallic 'species' in fact belong to the same species when cultures derived from their types are available.

Electrophoresis

Electrophoretic techniques provide a means of comparing the complex proteins in different organisms. Various methods are available but gel electrophoresis has proved of most value in the fungi. Starch or polyacrylamide gels up to 1 cm thick with a suitable buffer on glass plates are used. The extracts of the specimens being investigated are placed at one end and an electric potential of 6–10 volts/cm gel applied across the length of the plate. This causes their various ionizable proteins to move through the gel at different rates, so separating according to their ionic properties. The gel can then be stained for proteins and(or) particular enzymes when the run has been completed and the positions of the bands produced by the various extracts compared. For an introduction to electrophoretic techniques see Smith (1960).

Gel electrophoresis has already proved to be of value in some groups of fungi at the specific level (see e.g. Durbin, 1966; Pelletier and Hall, 1971; Reddy and Stathmann, 1972; and other references cited by Hall, 1969, and Norris, 1968) but does not appear to have been used in the lichens. It is clear that these techniques need to be employed more widely in the fungi than they have been so far but, as in all chemotaxonomical studies, it is essential that as many isolates as possible are examined to establish the limits of infraspecific variation in band patterns before using the data in any taxonomic discussions.

Host specificity

The specificity of fungi for particular host plants has assumed an important rôle in the taxonomy of many groups of microfungi which cannot easily be grown in culture. In some large genera identification is facilitated by first checking the species described from a particular host plant (see p. 49) but an unfortunate tendency to describe as new species material of a genus found on a new genus of host plants has arisen. In some cases clear morphological differences accompany the restriction to a particular host and so this approach may be justified, but in many instances the morphological differences may be of a minor nature or not apparent.

Where isolates may be grown in pure culture the rôle of the host in determining any morphological variations may be assessed, but where this is not possible attempts should be made to inoculate the fungus into other hosts, starting with allied species and genera. Where a fungus is found to attack only a particular host and is not distinguished morphologically it should be regarded as a special form or physiologic race rather than a distinct species.

Various methods of infecting hosts by fungi are available depending on the method of its dissemination and entry into the host (see Wheeler, 1968, for review), but a detailed discussion of these is outside the scope of this book. Further information may be obtained by reference to standard texts on plant pathogenic fungi (see Ainsworth, 1971) and issues of the *Review of Plant Pathology* (CMI, 1922→). Cross inoculation experiments need to be carefully designed to ensure that the resultant information is statistically reliable; for hints on experimental design see Commonwealth Mycological Institute (1968).

Numerical taxonomy

Numerical taxonomy (sometimes called 'taxometrics') aims '. . . to arrive at judgements of affinity based on multiple unweighted characters without the time and controversy which seems necessary at present for the maturation of taxonomic judgements' (Sokal and Sneath, 1963); i.e. it aims to eliminate observer bias from taxonomic schemes. Many different methods are now available but all involve computer techniques. These fall into two main categories: (1) divisive-monothetic methods which consider the entire population and split it up hierarchically on the basis of the presence and absence of particular characters in turn, and (2) agglomerative-polythetic methods which employ indices of similarity based on all possible characters to combine individuals which are most similar into groups. Methods of this latter type are of most value to the taxonomist as they produce a quantitative measure of phenetic similarity between the OTU's* (Ivimey-Cook, 1969). These relationships are worked out in a 'Q-matrix' which can be studied in one of two main ways:

(a) *Principal component analysis.* This method, which was not originally devised for taxonomic applications, extracts the principal axes from a multi-dimensional array of variables. These principal axes ('components') are taken in pairs and the OTU's plotted on them. The components are normally automatically calculated in decreasing order of variance (usually equivalent to taxonomic importance). When to cease extracting components and the significance to be given to the results from each pair remain, however, necessarily subjective steps.

(b) *Cluster analysis.* This approach is designed to sort out groups of OTU's the members of which have greater affinities to each other than they do to any other groups being studied. The 'pair-group' type of cluster analysis takes in turn each pair of OTU's; each pair is treated as a single unit and used to calculate new matrices. 'Variable-group' types consider each OTU in turn for addition to a cluster initial to form a larger cluster. The 'variable-group' methods depend on the selection of arbitrary criteria which must be satisfied before a particular OTU can be admitted to a cluster and for this reason the 'pair-group' approach has been considered as most suited to taxonomic use (Ivimey-Cook, 1969). The 'variable-group' method of Carmichael and Sneath (1969), however, overcomes this difficulty by limiting the sizes of clusters when the addition of the closest member would cause, for example, a large drop in the average linkage values between other members of the cluster, and repeating the process at different levels of resolution (i.e. with different values for single-linkage and average-linkage parameters). This method, which was designed primarily for taxonomic use, produces clusters, provides distances between OTU's within their clusters and between different clusters, also stating why particular OTU's have been excluded from individual clusters.

The results from numerical analysis are usually displayed in the form of either phenetic dendrograms or cluster maps.

The first step in carrying out a numerical study is to score the characters and character states of the OTU's in tabular form. The characters used must be those which are not, as far as can be determined, dependent on one another as this will tend to make the resultant data misleading. In addition to anatomical and morphological characters chemical components, cultural characteristics and physiological parameters may also be used where appropriate. As many characters as possible should be scored for each organism; about 60 is ideal but satisfactory results may be obtained with 30–40 in some cases.

The groups established by numerical methods must be analyzed carefully. Where the results differ from accepted classifications this usually seems to be due to insufficient characters being considered (see e.g. Stearn, 1968). The value

* OTU's = operational taxonomic units; i.e. the individual specimens or taxa (usually species or genera) being studied.

of numerical techniques is that one has to score taxa carefully for each character (which may also be valuable in other connections; see p. 73) and that the alignment of OTU's of uncertain affinity may be made more apparent. It provides good tests for taxonomic schemes which are based on more orthodox taxonomic approaches using weighted characters and consequently is of value in monographic studies.

For more detailed information on techniques see Sokal and Sneath (1963), Cole (1969), Lockhart and Liston (1970) and Sneath and Sokal (1973); and for examples of applications in the fungi see Campbell (1971) and references cited in Ainsworth (1971). A useful review of criticisms of the techniques is provided by Sneath (1971). The uses of computer methods in ecological work (p. 96), key construction (p. 74), flora compilation (p. 78) and herbaria (p. 31) have already been referred to.

Ontogeny

The way in which ascocarps are initiated and develop is proving to be of considerable taxonomic importance in both lichenized and non-lichenized Ascomycotina. Ascocarp ontogeny is studied by means of thin sections (usually 5 μm) prepared either by embedding in wax or with the aid of a freezing microtome (see p. 22) and staining with haematoxylin, eosin or cotton blue. Ascocarps at various stages of development are examined and camera-lucida (see p. 59) drawings prepared from them. The ontogeny of lichen thalli may also be studied in the same way.

Conidial ontogeny in Deuteromycotina may be investigated by means of time-lapse photomicrography (see e.g. Cole and Kendrick, 1968).

For examples of ontogenetic studies and their use in taxonomy see Bellemère (1967), Chadefaud and Avellanas (1967), Henssen (in Lamb, 1968), Henssen (1970), Jahns (1970), Jahns and Beltman (1973), Janex-Favre (1971), Letrouit-Galinou (1968, 1971) and Letrouit-Galinou and Lallemant (1971).

Serology (Immunology)

This technique, like that of gel-electrophoresis, is a means of comparing the proteins present in different specimens. If proteins which are sufficiently different from those which occur naturally in an animal's body (usually a rabbit is employed) are introduced into it, specific antibodies which inactivate the introduced proteins ('antigens') are produced in its blood. An antibody containing serum ('antiserum') prepared from the blood of the animal can be employed to produce visual serological reactions of two principal types, agglutination (clumping) and precipitin (precipitation from solution by titration). Antigens prepared from freshly collected material or living cultures (preferably in liquid media) can be tested against one or more antisera and some idea of the similarity of the protein components obtained.

Serological tests have proved extremely valuable in the taxonomy of bacteria and viruses but have been employed only occasionally in mycology. Agglutination tests have proved useful in the yeasts but precipitin reactions have been **found to be more suitable in mycelial fungi. Those studies which have been** carried out in the fungi indicate that serological methods can provide information of taxonomic value and comprehensive serological surveys consequently seem likely to prove to be of considerable importance. A brief introduction to the techniques is provided by Commonwealth Mycological Institute (1968) and reviews of their applications in mycology are provided by Proctor (1967) and Preece (1971a). The rôle of serology in yeast classification is discussed by Richards (1972). Agglutination reactions have also been demonstrated with lichen antigens (e.g. Barrett and Howe, 1968) but so far serological methods have been too little used in this group to assess their possible significance.

Some immunological reactions may be combined with other procedures. The fluorescent antibody technique (Preece, 1968, 1971b), for example, enables the hyphae of particular fungi to be identified in and on plants or soil without the need for culturing and so is valuable in studying host-parasite relationships etc. Precipitin reactions may also be used in conjunction with electrophoretic techniques ('immunoelectrophoresis') and this has proved of some value in taxonomic studies in vascular plants (e.g. Fairbrothers, 1968).

Transplants

Transplantation of intact lichen thalli into different habitats or areas affords a means of investigating both their tolerance of particular environmental conditions and the rôle of the environment in the determination of thallus form. Transplant experiments have been mainly used in ecological investigations, particularly in relation to the effects of air pollutants (see Hawksworth, 1973c), but may also provide data of taxonomic value (see Hawksworth, 1973b).

Specimens are either removed on their substrate (e.g. on bark disks) or free of the substrate and attached by a non-volatile resin glue (e.g. Araldite) and studied (usually by colour photography) at regular intervals. Because of the slow growth rates of lichens transplant experiments must be essentially long-term, but Richardson's (1967) study on the maritime morphotype of *Xanthoria parietina* demonstrates their value. Transplant techniques clearly deserve to be used more widely in lichen taxonomy than they have been so far.

Ultrastructure

The advent of electron microscopy has led to a greatly increased knowledge of the structure, composition and organelles of both the non-lichenized (Bracker, 1967) and lichenized (Jacobs and Ahmadjian, 1969) fungi. These techniques are also proving to be of value to the taxonomist as they enable structures which have not previously been known to exist or which it has not been possible to study properly by light microscopy to be used in taxonomic studies. Transmission electron microscopy has been found to be particularly useful in the examination of asci (see Griffiths, 1973), flagellate zoospores in the Mastigomycotina, and the surface features of ascospores and basidiospores (using carbon replica or freeze-etching techniques). The structure of conidiogenous cells and the precise mode of conidial ontogeny, which is now considered to be of paramount importance in the Coelomycetes and Hyphomycetes, can also be studied by electron micro-scopy (e.g. Campbell, 1972; Olá'h, 1972; Sutton and Sandhu, 1969). Material to be examined in section is fixed, embedded and sectioned on a special glass knife edge, while replicas are formed by coating the surface with carbon and shadowing with metal to accentuate the features. The most satisfactory way to learn the techniques of electron microscopy is to work for several weeks in a laboratory where these instruments are routinely employed. The preparation of mycological specimens for transmission electron microscopy is discussed by Greenhalgh and Evans (1971), who include a valuable list of references, and useful introductions to the handling of biological materials in electron microscopy are provided by Hayat (1970), Weakley (1972) and Koehler (1973).

The more recent development of the scanning electron microscope (for details of the theory see Oatley, 1972) has enabled mycologists to study specimens directly and in three dimensions without the need for any lengthy complex fixing and sectioning preparative stages. The specimens are attached by double-sided adhesive tape or a suitable glue to metal stubs, subjected to a vacuum, coated with a fine layer of metal (e.g. aluminium, gold/palladium), and examined directly in a vacuum by a scanning electron beam which produces an image on a screen. As the material is not destroyed it can be kept for future reference. In scanning electron microscopy only the surface features can usually be observed but it is these which are often of particular interest to the taxonomist. The scanning electron microscope has numerous applications in the fungi and lichens but its potential is only just coming to be realized. In the case of the fungi this instrument has proved of particular value in the study of the surface features of spores (see e.g. Pegler and Young, 1970), while in the lichens it is the surface features of the thalli themselves (e.g. Hawksworth, 1969b; Hale, 1972, 1973) which have proved of particular taxonomic interest and not only the ascospores (e.g. Tibell, 1971). Ornamentation on hair-like hyphae may also prove of considerable taxonomic interest when examined in this way (see Hawksworth and Wells, 1973). This microscope is also useful in that good quality habit photographs with deep fields in the range ×60–250 (which are difficult to achieve by conventional photomicrography) are readily obtainable (e.g. Plumb and Turner, 1972). The investigation of the anatomical structure of lichen tissues in cut portions of thalli is also facilitated (e.g. Hawksworth, 1972a; Hale, 1973) and it is clear that this will also be true for other fungal tissues.

If a taxonomist has access to a scanning electron microscope he should endeavour to use it in his studies as much as is practicable. A general account of the use of the scanning electron microscope is provided by Hearle et al. (1972).

LITERATURE

FINDING relevant literature is an important aspect of taxonomy; indeed before any taxonomic manuscript is submitted to an editor the author should have seen all works which might be pertinent. It is not the purpose of this section to provide a comprehensive summary of the vast literature of mycology, now an almost impossible task, but to indicate how relevant literature references can be traced and how, once they have been traced, copies can be located. D. J. de S. Price showed that the 'literature explosion' of science generally follows the exponential portion of a logistic growth curve with a doubling time of about fifteen years. This is also true for systematic mycology; the *Bibliography of Systematic Mycology*, for example, listed some 6,325 references during the ten-year period 1950–1959, but 11,377 during the period 1960–1969. Ainsworth (1971) estimated that the current annual output of all mycological literature was of the order of 5,000 items dispersed through not less than 3,000 books and journals. Brookes (1971) considered the 'information explosion' less alarming as it is counterbalanced by discarding of the old, but this is not true for the taxonomist who frequently has to consult literature a century or more old. It is therefore essential that a mycologist should know just how to find what he ought to see in the minimum of time. For a general account of biological literature and its use see Bottle and Wyatt (1966).

Sources of references

General textbooks, taxonomic works, floras and other geographical works, bibliographies, abstracting journals, periodicals, and major works on each genus are listed in Ainsworth (1971). This book provides the starting point for finding literature on all aspects of mycology and lichenology. While this work attempts to be as complete as possible only major works are cited so the absence of any monograph under a generic name does not indicate that no information is extant. From the references in Ainsworth (1971) it is usually possible to trace much of the earlier literature through the bibliographies in those works. In the only reference cited under *Podospora* Ces., for example, nineteen more are given, and one of those has a bibliography of 164 references. The mycologist interested in a genus is therefore able to compile a fairly comprehensive list of references by working backwards through the literature in this way.

Short papers particularly are likely to be missed by this method but might be picked up by others. Many taxonomic papers contain new taxa and can be traced through indices of names (see p. 108) and similar publications. If a considerable number of new taxa in a group are mentioned it is possible that the work cited may be a monograph or revision. All important taxonomic papers do not contain new taxa, however, and so would not be located in this way but these may often be traced through bibliographic and abstracting journals (see Ainsworth, 1971; Commonwealth Mycological Institute, 1971).

The major abstracting journals of value to the mycologist are *Abstracts of Mycology* (1967→) [*Mycological Abstracts*; reprinted from *Biological Abstracts*], *Bibliography of Systematic Mycology* (1943→), *Excerpta Botanica* (1959→) and *Microbiology Abstracts, sect.* C (1972→). For information on pathological aspects of fungi the *Review of Plant Pathology* (1922→) [formerly the *Review of Applied Mycology*] and the *Review of Medical and Veterinary Mycology* (1943→) should be consulted. The following works should also be scanned from time to time: *Asher's Guide to Botanical Periodicals* (1973→), *Bulletin Signalétique, sect.* 14 (1940→), *Current Contents* (1959→), *Dissertation Abstracts* (1938→), *Fortschritte der Botanik* (1931→), *Index of Conference Proceedings received by the NLL* (1964→), *Index to Theses accepted for higher degrees in the Universities of Great Britain and Ireland* (1950→) and *Referativnyĭ Zhurnal* (1954→). The services of the *Science Citation Index* (1961→) may also be of value in some cases.

Lichenological literature is also listed by W. L. Culberson (1951→) and in the *Bulletin of the British Lichen Society* (1958→) and the *Revue bryologique et lichénologique* (1928→). Specialist literature lists are featured in newsletters, such as the *Neurospora Newsletter* (1962→); a list of newsletters is included in Commonwealth Mycological Institute (1971). Booksellers' and publishers' lists often contain information on books in press. Edwards (1971a) summarizes the available abstracting and indexing services in the biological sciences.

If a particular author is known to have been very interested in the group on which the literature is required all his works should be studied as they may contain footnotes and asides which may be relevant although their title does not mention the particular group by name; this is particularly important in geographical lists and floras, titles of which often omit to mention the taxa covered.

A great deal of pertinent literature may be discovered simply by browsing amongst library shelves and studying the incoming books and journals which are displayed in major libraries. Indeed, if the mycologist has access to a large mycological library, time devoted to looking through reprint boxes, bound collections of reprints, and volumes on the shelves will often prove rewarding. It is often possible to pick up items missed by previous workers in this way. 'Vascular plant' floras, particularly those published last century, should not be passed over rapidly either as they often contain appendices dealing with lichens and fungi which are sometimes by different authors.

Indices to journals such as the *Review of Plant Pathology* (1922→), *Bryologist* (1898→), *Mycologia* (1909→) and the *Transactions of the British Mycological Society* (1896→) should also be studied as many useful data are often readily obtainable in this way.

One of the most valuable sources of literature references is, however, someone who has studied a group in detail. Many specialists build up their own card indices to relevant literature and some are willing to place these data at the disposal of other workers. They may also know of recent publications which have not yet had time to be processed in bibliographic and abstracting journals or of earlier papers which have been overlooked. If the opportunity arises the student should therefore endeavour to enlist the help of known specialists in his search.

Catalogues of names

Lists of names of taxa of fungi and lichens together with their places of publication are given in the following major works:

Cooke and Hawksworth (1970) [fungi and lichen family names]
Index of Fungi (CMI, 1940→) [fungi 1940→; lichens 1970→]
    Supplement: A supplement to Petrak's Lists 1920–1939 (CMI, 1969) [fungi]
    Supplement: Lichens 1961–1969 (CMI, 1972) [lichens]
Lamb (1963) [lichens]
Petrak (1930–44, 1950) [fungi; lists I-VII (1930–44) reprinted by CMI, 1952–55]

Saccardo (1882–1931, 1972) [fungi]
Zahlbruckner (1921–40) [lichens]

Taxonomists will also have to refer to lists of generic names in other plant groups when searching for possible homonyms. Most are catalogued in the following works (see also Leenhouts, 1968):

Dawson (1962) [algae]
Genera Filicum (Copeland, 1947) [ferns]
Index Bergeyana (Buchanan *et al.*, 1966) [actinomycetes; bacteria]
Index Hepaticarum (Bonner, 1962→) [liverworts]
Index Filicum (Christensen, 1906–65) [ferns]
Index Kewensis (Hooker and Jackson, 1895→) [flowering plants]
Index Nominum Genericorum (1955→) [card index; will eventually cover
    names in all plant groups; see Cowan (1970).]
Index Muscorum (van der Wijk *et al.*, 1959–69) [mosses]
Kylin (1954) [red algae]
Mills (1933–35) [diatoms]
Sylloge Algarum (De Toni, 1889–1924) [algae]
Vanlangingham (1967→) [diatoms]
Willis (1973) [vascular plant genera]

Tracing incomplete and 'incorrect' references

In published papers phrases such as 'Arnold considered this might be a species of *Lecidea*', without any reference to any paper by 'Arnold', will be encountered from time to time. It is important for a monographer to locate references such as this and find out what 'Arnold' actually said and why. Another common practice, particularly last century, was merely to mention the author and title of an article even though it appeared in a journal. Much time can be wasted searching for a 'book' which in reality was an article in a journal.

In instances such as these the key to the correct reference is the author's name. For the pre-1930 literature the compendia of Lindau and Sydow (1908–18) and Ciferri (1957–60) are invaluable. The literature on lichens to 1870 is also reviewed by Krempelhuber (1867, 1869, 1872) and the forthcoming work of Grummann (1974) may also be valuable in this connection. If these fail the article may be listed in the Royal Society of London's *Catalogue of Scientific Papers* (*1800–1900*) (London, 19 vols., 1867–1925), or the *Catalogue of the Books, Manuscripts, Maps and Drawings in the British Museum* (*Natural History*) (London, 5 vols., 1903–15; and *Supplement*, 3 vols., 1922–40) or the *Catalogue of the Printed Books and Pamphlets in the library of the Linnean Society of London* (London, 1925). For literature published after 1940 the bibliographic and abstracting journals mentioned above (pp. 107–108) should be searched. Obituaries sometimes include full lists of publications of the deceased author and are consequently also valuable in such cases. The most complete list of obituaries available is that of Barnhart (1965); it includes most mycological and lichenological taxonomists (see p. 179). If the dates of birth and death of the author (see pp. 180–187) or the date of the publication (e.g. if the original phrase which provided the reason for the search was 'In 1885 Nylander . . .') are known this limits the number of catalogues to be searched. There is little point in looking through a catalogue of literature published in 1918 for an author born in 1930.

As a last resort collections of reprints in large libraries can be examined. The collected works of some authors have also been reprinted and issued in a single volume or volumes.

Problems often arise with periodical abbreviations which may apply to more than a single journal (e.g. '*J. Bot.*' and '*Mag. Bot.*') and in ascertaining which is the correct one the dates of the start (or end) of the journal are useful as is the correlation between volume number and year of publication. Some journals also have subtitles which are not always indexed by librarians and if the problem is finding the correct journal the assistance of someone with a knowledge of specialist scientific journals becomes necessary.

Incorrect volume numbers or page numbers are common mistakes often perpetuated through the literature as authors are prone to cite synonyms without ever seeing the paper concerned. In such cases it is advisable to visit a library with a complete run of the journal and check adjoining volumes and permutations of the page and volume numbers, dates etc.

Dates of publication

The date on which a particular work was published is very important in nomenclature (see e.g. pp. 129–151) and must be ascertained correctly. For early books the most valuable reference works are those of Pritzel (1871–77) and Stafleu (1967), which deal with the 'classical' literature about which there is often some controversy over the dates when particular publications appeared. Dates of publication of many taxonomic works are also listed by van Steenis-Kruseman and Stearn (1954).

Many large works appeared in parts (fascicles) over a number of years but are often bound together with a cover which gives the date on which the last part (and(or) index) appeared. Rehm's volume on the Hysteriales and Discomycetes in *Rabenhorst's Kryptogamen-Flora*, for example, was issued in 21 parts between 1887 and 1896. Fortunately the dates in many such publications are printed separately at the ends of such works and many institutions take care to bind them with their original wrappers.

Libraries also frequently stamp their books with the date on which they were received. While this is in most cases the same year or after the date on the work, in a few it is the year before that printed on it, indicating that it was issued earlier than its title page suggests.

Journals (particularly the last number for a particular year) often appear later than scheduled so may be dated earlier than they actually appeared. Most taxonomic journals now give the precise date of publication in the succeeding issue or list them together at the end of each volume. Authors should always make sure they check this carefully. It is a good practice when dealing with journals to assume the date is wrong and try and find an indication in later issues of when it appeared! If no such evidence is forthcoming or the library received it in the same year it may usually then be taken that the date on the work is correct. However, reprints were sometimes issued before the journal part in which they appeared (see p. 151) and so dates on reprints should also be checked when these are available.

Detailed lists of the dates of publication of some books and journals have been prepared, such as those of Raphael (1970) on the issues of the *Transactions of the Linnean Society of London* published between 1791 and 1875 and Stafleu (1972) on *Die naturlichen Pflanzenfamilien*. When such studies have been carried out librarians should be urged to write or type slips bearing the correct dates and attach them to the inside covers of each volume. Such slips should always include the source of the information.

Exsiccata sometimes have validly published names and so their dates must also be checked carefully. No comprehensive list for the fungi is available but Sayre (1969) deals with general cryptogamic and lichen exsiccatae, and Stevenson (1971) with North American fungal exsiccata.

Contemporary papers, reviews and journals listing abstracts of papers are also valuable aids in ascertaining correct dates of publication of particular works (e.g. see Stafleu, 1967).

Location

Having found a reference the work itself or a copy of it must be seen. Unless the student has access to a major library many works on systematic mycology will probably not be readily available to him. Just how to set about tracing copies varies in different countries according to the organization of their national library services, and the account which follows is intended principally for the guidance of workers in Britain and others wishing to obtain photocopies or loans from British libraries.

When the student has made certain that the work he requires is not in his own institution's library he should give his librarian full particulars of it—many libraries have forms designed for this purpose. The location of journals usually presents little difficulty if the correct complete title is known. The *World List of Scientific Periodicals Published in the years 1900–1960* (London, Butterworths, Ed. 4, 3 vols., 1963–65; see also supplements prepared from the *British Union Catalogue of Periodicals*, London, Butterworths, 1964→) gives the exact holdings of journals in all major libraries in the British Isles and information as to the completeness of their sets. The British Museum (Natural History) has the largest holding of natural history journals in the country, and the catalogue of them, *List of Serial Publications in the British Museum (Natural History) Library* (London, British Museum (Natural History) Publication no. 673, 1968) has some 12,500 entries. This library does not lend books or journals but will supply photocopies for individual scientific study subject to the laws of copyright. The National Lending Library for Science and Technology, however, lends volumes to approved libraries and so is valuable when precise page numbers of articles are not known; for a list of the current periodicals they take see *Current Serials received by the NLL March 1971* (London, HMSO, 1971).

Books present greater problems than journals as no comprehensive catalogue of the holdings of different libraries is available. To trace the location of early works the catalogues of the British Museum (Natural History) and the Linnean Society of London (see p. 109 above) will show if those institutions have the work. The Royal Botanic Gardens, Kew, and the CMI have between them the most complete set of mycological books available in the British Isles, but do not now* issue a published catalogue of them. If possible the student should visit these libraries and study the books he wishes to see; if copies of particular pages are required xerox copies can be provided subject to the laws of copyright (U.K. Copyright Act, 1956, and the Copyright Libraries Regulations, 1957).

The period of copyright is in most cases the life of the author and fifty years from the end of the calendar year in which he died. In the case of joint authorship this refers to the author who dies last. For articles from periodicals still in copyright single copies can be provided for research purposes; if more than one copy is required written permission from the copyright owner is required. Whole books, extracts from books or pamphlets, and more than one article from a single issue of a journal still in copyright cannot be copied unless permission has been obtained. Librarians generally require students requesting copies to sign copyright declarations to the effect (a) that they have not had a copy of the article from any library before, and (b) that the copy is required for their personal use. Once a copy has been obtained the making of a further copy from it without permission of the copyright owner may be an infringement of copyright.

Some difficulty may be encountered in xerox-copying thick works with over-tight bindings and microfilming of these may be necessary to obtain a satis-factory result. Half-tone plates often cannot be reproduced satisfactorily by xerography and microfilming or true photocopies are sometimes necessary to obtain adequate reproductions of these.

Students often request single pages of books or articles which bear the accounts of particular taxa but if the work has not been seen it is advisable to see it first to check that important information given in any introduction is not missed (e.g. what rank an asterisk denotes, where the type specimens are) and that plates and figures are not overlooked. It is useful when requesting a copy of a page of a book to obtain also a copy of the title page to ensure that it will be cited correctly. As in the case of journals, requests should usually be made through the librarian of the student's institution. The Inter-Library Loans system operated by the British Universities is often successful in obtaining obscure works with the assistance of the National Central Library, London (integrated with the National Lending Library in 1971) which has contacts with many libraries throughout the world.

Fortunately a great many mycological works are now available as microfiches prepared by organisations such as the International Documentation Company (IDC), Zug, Switzerland; these are read with special readers or a binocular microscope can be used; they can also be printed photographically and enlarged. Available microfiche editions are listed in *Taxon* from time to time and cata-logues of them are available on request.

* For works published prior to 1916 in K see the *Catalogue of the Library of the Royal Botanic Gardens, Kew* (*Kew Bull., add. ser.* 3:1–790, 1899) and *Supplement* (*op. cit.* 3 (2): 1–433, 1919).

Where there is a particular difficulty in obtaining a copy of a mycological work the librarians of taxonomic institutions (e.g. British Museum (Natural History), London; or CMI) should be approached for advice and assistance.

Libraries

Most major European and North American libraries contain a high proportion of the mycological works likely to be required by the systematist. A list of the principal libraries of the world is given by Esdaile (1957), and Edwards (1971b) gives a valuable account of British biological libraries and their scope. Many important mycological libraries belong to smaller institutions, however, such as the Farlow Reference Library of Harvard University, CMI and the Rijksherbarium in Leiden, or were built up by individual mycologists, such as those of H. A. Kelly (see Krieger, 1924) and A. H. R. Buller (see Oliver, 1965). Some scientific societies dealing with mycology have their own libraries which are available to members (e.g. British Mycological Society, British Lichen Society). The main British libraries of interest to the mycological taxonomist have already been referred to above.

Most taxonomists will find that they have to visit and use large libraries and if they do not have access to one normally it is important that the way to find the data required as quickly as possible on visits to such libraries is known.

The arrangement of books in different libraries varies so it is essential that the student should ascertain the arrangement employed before using them. Books and periodicals are usually arranged separately. The most widely used system of cataloguing used in British Libraries is the decimal system of M. Dewey first introduced in 1873 and now expanded and generally known as the Universal Decimal Classification (UDC) (British Standards Institution no. 1000A, 1961). The main numbers of interest to mycologists are:

5	Pure science
58	Botany
581	General botany
581.2	Plant diseases. Phytopathology
581.9	Distribution of plants. Phytogeography, Floras
582.24	Myxomycetes s.l.
582.28	Mycology. Eumycetes
582.29	Lichenes
616.992	Plant parasite infections. Mycoses
632.4	Parasitic diseases. Fungi and mould diseases
663.1	Industrial microbiology and mycology
664.642	Yeast and leaven

Some mycological works do not fall clearly into the above categories: mycological floras, for example, may be placed under 581.9 or 582.28; and systematic studies on plant pathogenic fungi under 581.2, 582.28 or 616.992. The UDC and the alternative Library of Congress System, are not suited to very specialized libraries as too few headings are used and such libraries (e.g. CMI) consequently develop variants more suited to their particular needs. Most libraries maintain an accurate card index which gives the precise location of microfiches, microfilms, pamphlets, books, periodicals and reprints and so if the student is searching for a particular work this should present no great problem. If a work seems to be wrongly classified this should be drawn to the attention of the librarian in charge.

Journals are most satisfactorily arranged according to their country of origin and again most libraries will have some index to the shelves on which they are located. Microfilms and microfiches are usually stored separately in filing cabinets. Collections of offprints are valuable parts of all botanical libraries and should also be carefully checked when searching for particular works. In some instances these are bound into 'opusculae' either by subject or by author and when this occurs they are often not catalogued comprehensively.

Before entering a library it is advisable to write each reference to be checked on a separate slip of paper or card. Books and journals can then be arranged alphabetically by author and title, respectively, to facilitate searching in the catalogue and should then be sorted according to their shelf-marks to save time walking from one part of the library to another.

When a volume has been taken from the shelf it should either be returned to the precise place it came from on the shelf or be given to the librarian as if it is returned to the wrong place it may not be refound by others requiring it. If any typographical errors, information on dates of publication, or other items which might usefully be noted in the volume are found these should be drawn to the attention of the librarian who can add an explanatory note on a label inside the cover; a user should not mark any work himself.

In using older works great care must always be taken to ensure that they are not damaged, for many are virtually irreplaceable and extremely valuable. They should, for example, be removed from the shelves carefully, not by pulling on the binding at the top of the spine; never bent backwards to keep them open; and open books should never be placed on top of one another. If some pages are uncut the librarian should be asked to attend to this.

Citation

The citation of works in the text (p. 118), and in lists of synonyms (p. 67) is referred to elsewhere in this book, and this section presents some general points to be borne in mind in citing any mycological work.

An accepted system should always be followed and adopted consistently throughout a particular article. Periodical abbreviations in Britain now generally follow those proposed in the *World List of Scientific Periodicals* ... and its supplements (see p. 111); these abbreviations follow the order of words on the title pages. A list of those most commonly used by biologists has been prepared by Williams (1968) and includes those most frequently cited in mycological publications. Some journals, particularly in North America, follow the principles of the 'Anglo-American Cataloguing Code' (Committee of the Library Association and Committee of the American Library Association, 1965; Committee on Form and Style of the Council of Biology Editors, 1972) in which the words of journal titles appear in a different order. The *British Standard on Bibliographical*

Citation (British Standards Institution no. 1629, 1950, and amendment issued May 1951) is used by some British journals. A few journals no longer use abbreviations but give the titles in full.

If a periodical to be cited is not listed in, for example, the '*World List . . .*' its title should either be given in full or be abbreviated according to the same principles. If the author prefers to use a subtitle or alternative title he should only do this if both titles are cross-referenced in the '*World List . . .*'. In citing journal articles in lists of references and bibliographies the precise title of the article, date, volume (and part number where each part has an independent pagination) and first and last page numbers should all be given. It is often useful to place journal title abbreviations in italic type, the volume number in bold type, and part numbers in parentheses in normal face type after the volume number.

There is no universal system for abbreviating the titles of mycological books but recommended abbreviations for some of the principle works commonly cited in synonymy are listed on pp. 193–198. It is often tempting to refer to large papers issued in journals by abbreviated titles of the paper instead of the abbreviated title to the periodical where they appeared. This method, which was common practice last century, should not be adopted, however, as it may prevent someone locating the work in a library which has the journal but not a separately bound offprint (and consequently no card for it in its catalogue). Book title abbreviations are only used in citing places of publication of names in taxonomic accounts. In citing books in bibliographies it is usual to place the title in full in italic type and the place of publication (and publisher) in normal type but some journals depart from this procedure.

In lists of references in papers using an alphabetical system (see p. 118) authors are best entered surname first followed by initials of the forename(s) although some journals place the authors' initial(s) first. Compound surnames are a continuing source of ambiguity and confusion. If a name is hyphenated enter it under the first part of the name (e.g. Müller-Argoviensis, J. *not* Argoviensis, J. M.-). In some cases it may be best to follow the author's own or accepted usage of the form for a particular name. The custom of the authors' country of origin is always adopted; e.g. the Chinese 'Lu Ding-An' is listed as 'Lu, Ding-An' *not* 'Ding-An, L.' and the Portuguese 'C. Castelo Branco' as 'Castelo Branco, C.' Prefixes compounded with names are retained (e.g. Vanderhoek, McAlpine) but where the prefix is separate the name is entered under the part following the prefix except in French, Italian and Spanish where the prefix consists of an article, e.g.

Höhnel, F. H. von (German)	De Sloover, J. R. (French)
Overeem, H. van* (Dutch)	De Notaris, G. (Italian)
Smet, F. de (Flemish)	La Farina, M. (Spanish)

It is not possible to discuss the citation of all types of personal names in various languages here. The national usages of names of authors in all major countries

* 'Van' is usually indexed in English publications but not in Dutch ones and so has been retained in the references in this book.

for entry into catalogues and lists of references are summarized in the comprehensive manual by Chaplin (1967).

If a work appeared in a book edited by another author the editor's name, title of the whole volume, pagination of the particular chapter, and place of publication and the name of the publisher require citation in addition to the author's name and the title of the paper or chapter being referred to. In the case of works agreed upon, authorized or issued corporately by a Government Department, Institute, Society or other organization without mention of a particular author, editor or compiler, the organization is considered to be the author (i.e. 'corporate authorship'); e.g. Commonwealth Mycological Institute (1971).

For examples of the form of citation of references see the list of those referred to in this book (pp. 208–221). Citations in synonymy present special problems and are referred to on p. 66. For details of methods of referring to particular works in the text of published papers see p. 118.

PUBLISHING

WHEN important new work has been undertaken the mycologist will wish to make his information available to as wide an audience as possible; i.e. he will want to publish it. Notes on the particular problems associated with the preparation and publication of monographs, revisions and floras and the types of data which should be included in them have already been discussed. It is the aim of this section to make some general points pertinent to the preparation of any mycological manuscript for publication and to present notes on the checking of manuscripts and printer's proofs.

The enthusiastic worker who makes what he believes to be a new discovery, of, for example, a new species, may be tempted to publish it immediately before someone else discovers it. He may also believe that *any* publication will help him to obtain a scientific position or increase the respect of his colleagues for him. Such is not always the case in practice, however, and the following points should be borne in mind before proceeding to publish an article: (1) the scientist is primarily concerned with the advancement of Science and not of himself; (2) further collections or study might reveal important additional data and forestall the publication of erroneous information; (3) a single comprehensive paper is often preferable to numerous short notes and that the literature of mycology should not be unnecessarily inflated; (4) has the work been studied and checked by other specialists and considered to merit publication by them; (5) a poor paper might do more harm to a mycologist's reputation than good as it will undoubtedly be criticized in print in the future; and (6) it is much easier to make a scientific 'discovery' than to learn whether it has already been made.

On the other hand students should not go on accumulating data indefinitely without making it available to other workers. A great deal of valuable information has never reached the literature because of this. A happy medium should therefore be aimed at.

Although a single comprehensive paper is preferable to numerous brief notes this does not mean that there is never a need for short notes. Such notes can either provide an introduction to work in progress, present preliminary results and publicize the fact that one is working on a particular topic; or give all the available data on a particular small point out of the main field of one's research that would appear out of place or be likely to be overlooked in a longer paper. The former type of note may lead to difficulties if shown to be in error by subsequent work and so should be avoided. Newsletters and bulletins provide a very important medium for reporting work in progress and preliminary results and should be used where possible for brief notes of this type. Some journals specialize in short notes (e.g. *Nature, Lond.*) and are important as they provide a more rapid outlet than many specialist journals which appear much less frequently. By its very nature, however, systematic mycology only very rarely lends itself to notes of this type.

Manuscripts

Before starting to write a paper one should think it out carefully. The precise layout will be determined to some extent by the journal to which it is hoped to submit it, and the best guide to this is the 'Notes for contributors' which appear in most journals (frequently on the inside of the back cover) and the form of papers in recent issues. The manuscript itself should be clear, precise, logical and, as far as is feasible, brief. Titles must be short and their phrasing clear so that anyone scanning the contents list of a periodical is unlikely to miss anything of interest. Open-ended series titles like 'Mycological notes no. 35' are not helpful and should be avoided. If numbered series are to be employed they should be finite and each should have a clear individual title. The abstract (synopsis or summary) is the only part of a paper that many mycologists will read and it is therefore essential that this does not omit anything important. Abstracts should not normally exceed about 200 words and should mention specifically any new taxa described or new combinations proposed.

The address of the author and acknowledgements for any advice received from colleagues and specialists should be included. If an author has recently moved from the institution where the work was conducted his new address can be indicated in a footnote. Footnotes should, however, be avoided where possible as they are liable to be overlooked and break up the continuity of a paper. Any drawings, maps or photographs should be directly relevant and clearly show what they are labelled as illustrating; all too often published illustrations are misleading. The preparation of illustrations has been discussed on pp. 57–62. The place where an illustration is to appear is indicated in the text in the following way:

> Fig. 5 near here

The language in which the paper is written should be that of most of the people who will wish to read it. In the case of floras, particularly those designed for identification, this will be that of the country with which it is concerned, but in the case of systematic work such as monographs English or Latin is preferable.

Bourne (1962) notes that in the biological sciences 39% of the literature is in English, 13% in German, 13% in French and 8% in Slavonic languages and these proportions appear to be about correct for mycology. It is also the practice of many authors writing in languages not familiar to most scientists, such as Japanese and Polish, to include extended summaries in English or another European language. Some journals now include abstracts in several different languages as a matter of course.

The style should be as lucid as possible and should avoid unnecessarily abstruse terms and long sentences. Experience has shown that clarity is best achieved by fairly short sentences and that long sentences lead to ambiguity. Where special terms are employed their definition should follow an accepted system (e.g. Ainsworth, 1971; see also p. 70). The *Concise Oxford Dictionary of Current English* (Ed. 5, Oxford, Clarendon Press, 1964) is used as the standard for spelling in most journals but several other works will be found helpful for usage and style; e.g. Fowler (1965), Collins (1956), Hart (1952) and Turner (1964). The Royal Society of London's booklet *General Notes on the Preparation of Scientific Papers* (London, 1965) provides a valuable general introduction to the writing of scientific papers and articles. A good style comes only with practice but care and many re-writes of the first manuscripts an author produces will help him towards this end. Reading papers by established mycologists will also yield valuable suggestions, as will detailed criticism of other published papers from this standpoint.

References which are relevant should always be cited. Papers should not be mentioned merely because the author knows of them; there is no virtue in a lengthy bibliography if the works cited in it are not directly pertinent. All references, particularly in taxonomic papers, must be checked in the original with particular attention to the precise title, volume (and where appropriate part) number, page numbers and the date of publication (see p. 110-111). Where particular works are stated to give certain information the author should re-check what the paper said. Some instances are known where authors have cited the wrong paper by a particular author published in the same year as the one they intended to cite! Careful checking must be made of the final manuscript to ensure that all the references are in fact in the bibliography; all too often they are not. If for any reason a paper has not been checked 'Not seen' (or '*non vide*', or '*n.v.*' in synonymy) should follow the citation and 'Cited by' or '*fide*' added after the reference; a conscientious author will not wish to be credited with an earlier worker's slip. In the text references are usually cited in the Harvard form; i.e. Smith (1921), Smith (1921, p. 154) or Smith (1921: 154), and (Smith, 1921); or by the use of numbers inserted in the text (e.g. (3) or [3]). In the former references are listed alphabetically at the end of the paper whilst in the latter they are placed in order of citation. The Harvard system is preferred as a specialist reading the paper may immediately know the work of 'Smith (1921)' when he sees it in the text without having to refer to the bibliography. The citation of books and papers in lists of references is discussed on pp. 114-116.

If any parts of the text are required to appear in bold type this is indicated by underlining with a wavy line (〰〰〰); words or figures to be placed in italic are underlined with a single line (————); and any to be placed in small capitals underlined twice (▭▭▭▭). Marking up of title, sectional and other headings should be left to the editor.

When the first draft has been prepared it must be examined very critically. The author himself may find slips and omissions by looking at it from the standpoint of a critic and discover points in need of further work before proceeding further. When the author himself is satisfied it is a good idea to submit it to colleagues who are not specialists in his precise field but who may be able to point out sections which are not quite clear; and to specialists working in the same field for detailed criticism. As much advice as possible should be obtained at this stage. When as much help has been received as is possible the author will be able to incorporate it and produce a further version of the manuscript which he may like to submit to those who read the earlier version to ensure that the points they raised have been adequately dealt with. Often it will prove necessary to rewrite the paper several times and in the course of rewriting particular care must be taken to ensure that errors (e.g. wrong page numbers and measurements) are not introduced. It is a good policy to re-check all numerals and references in the final version of the manuscript before submitting it to an editor.

The paper should then be put aside for some time (several months to one year is recommended) and then brought out and reconsidered. If the author still thinks it worthy of publication and is still satisfied with its style it is time to think of sending it to an editor. Manuscripts should be typed on one side only on good quality white paper of a reasonable size (e.g. foolscap, A4) with adequate margins all round (c. 3 cm). Many journals ask for more than one copy to be submitted but even where this is not stipulated most editors find it valuable. The author must, of course, retain one copy for himself as this will be necessary for checking proofs and is a safeguard against loss in the post, by a printer, editor or referee (all are known!).

Which journal?

The most suitable journal to submit a manuscript to is that which will be seen by most of the readers who will be interested in the paper. For local floras, additions to floras (e.g. 'First records of five Pyrenomycetes from Pakistan'), or other work primarily of local interest, local and national journals are obviously more suitable than international specialist journals which local amateurs and other non-specialists are unlikely to see. Papers which are monographs, revisions or nomenclatural notes transcend national boundaries (unless they are revisions of taxa in a particular geographical region) and should consequently be published in journals which will be read by specialists throughout the world. Taxonomists should also bear in mind the conditions for effective publication outlined in the Code (Art. 29; see p. 151).

If for any reason a paper is not accepted by a journal to which it was submitted the author should bear in mind the editor's and(or) referees' reasons for rejecting it and if necessary rewrite the paper or submit it to a more appropriate journal recommended by them. They should avoid the temptation when they have had a paper rejected by a specialist journal of submitting it in the same form to a non-specialist journal whose editor, not being a specialist, may accept it because he is unable to detect the deficiencies noted by the specialist editor.

Specialists to whom the manuscript was submitted for comment before sending it to a journal will often advise on the most appropriate place of publication.

Editors and referees

When a paper is sent to the editor of a particular journal this will be taken by the editor to imply that it is original work of the author or authors and that it has not been submitted to another journal. There has been a tendency, particularly in recent years in the physical sciences, for authors to publish the same article in two or more journals (see Anon., 1971) but this practice is to be deplored. The manuscript which arrives on the editor's desk should be in its final fully corrected form. If an author finds an error after submission he should inform the editor as soon as possible so that it may be corrected before the paper reaches a printer. The editor will ensure that the style of the paper is similar to that adopted in his journal and will often check as much of it as he is able (e.g. the references). In many cases the editor will not feel competent to criticize a paper in detail himself if it is outside his own particular field and will submit it to an assistant editor or referee with specialist knowledge for comment.

The referee will study the paper in detail, ensure that the work is in fact new and that the previous work has been adequately summarized, and return it to the editor with any comments and suggestions for improvement. If the comments are minor the editor will then send the paper to the printer and inform the author that it has been accepted. If any more important points have come to light the manuscript will be returned with these comments. In some instances a paper will be returned simply because it was sent to an inappropriate journal. The author should study in detail any comments made by editors and referees as these will certainly require attention before the paper is re-submitted. He should not simply deal with those he feels need attention and return the manuscript saying all have been dealt with as the editor may not study the paper in such detail again. If he disagrees with any point raised the author must explain this to the editor.

Editors of most mycological and lichenological journals are not professional (paid) editors and undertake the work in their spare time on behalf of the Society or other organization sponsoring the journal. The work they have time to do is consequently limited and editors naturally favour authors whose typescripts are perfect when they receive them because they do not have time to devote to extensive editing of particular articles. Authors should do all in their power to make the editor's task as easy as possible.

Proofs

Printers' proofs will be sent to the author by the editor as soon as he receives them, usually with the original manuscript. Proofs may be of one of two types: (a) galley-proof which is not set into page-form, and (b) page-proof which is in the form of the pages which will be printed. Only one set of proofs, which one depending on the journal, will normally be sent. When an author receives his proof he must check it very carefully word by word and figure by figure against his manuscript. This is the last chance he has of correcting any slips. Some authors find it helpful to have colleagues check the proof against their manuscript as well, as this reduces the risk of missing mistakes made by the printer. Any errors noted should be clearly marked according to the symbols adopted by the British Standards Institution (1958). Those which are most commonly needed in systematic papers are shown in Fig. 21. It is often helpful to indicate errors which are the printers in red and those which are corrections of the author (i.e. not in the original manuscript) in blue or black as this facilitates charging the author for corrections which are due to his mistakes. Any incorrect use of symbols can be misleading and may lead to the printer altering something incorrectly. Where a complicated change has been made it is often helpful if the author attaches a retyped correct version to the proof. Any directions intended for the printer and not to be set in type are encircled. If a word or phrase is deleted by the author it should be replaced with a word or phrase of the same length where possible to minimize type resetting costs. Remember not to make changes unnecessarily; the insertion of a comma might be as costly as the replacement of a whole line.

Proofs should be returned as rapidly as possible. Some journals ask for them by return of post and authors should endeavour to comply with the journal's instructions. If there is a considerable delay the paper may be sent back to the printer and published without the author returning the proof, resulting in the inclusion of slips not noted by the editor.

If for any reason a considerable number of changes are made in proof and the author is concerned about errors which might be introduced in resetting he should request a second set of proofs from the editor.

Reprints

Reprints ('separates' or 'offprints') may usually be ordered when the corrected proofs are returned to the editor. The number ordered must be final as when the journal and reprints have been printed the type is usually broken up by the printer.

The number of journals is very large and because of this most individuals and smaller institutions will only be able to subscribe to relatively few. Reprints are consequently one of the best ways available to an author for making his work known. In addition to his colleagues and people who have helped in the preparation of the paper he should endeavour to send copies to (a) specialists working in that field; (b) editors of bibliographic, abstracting or indexing journals (e.g., the *Index of Fungi* and *Bibliography of Systematic Mycology* for systematic papers); (c) the principal mycological institutions and herbaria throughout the world; (d) all herbaria cited in the text; and (e) the authors of any recent papers cited in it. In general the better known and more widely read the journal is, the less is the need to distribute reprints.

The individuals to whom an author sends reprints will often make a note of his address and add him to their mailing list for their own publications and so

Marginal mark	Meaning	Text mark	Corrected
ℰ	delete	ſspecies	species
ℰ	delete and close up	nommenclature	nomenclature
s/	replace by letter(s) indicated	syʏtematic	systematic
o/	insert letter(s) indicated	taxonmy	taxonomy
#/	insert space	thegenus	the genus
ital.	put in italic type	Amanit(opsis)	*Amanitopsis*
bold	put in bold type	(sp.nov.)	**sp.nov.**
rom.	put in Roman type	ascosp(*ores*)	ascospores
trs	transpose	con|ophores|idi|	conidiophores
l.c.	put in lower case	(SPECIMENS)	specimens
cap.	put as capital	see (f)ries	see Fries
wf.	wrong fount	Ar(ti)cle	Article
═	align	Reco^mmen^dation	Recommendation
⌒	close up	morph ͡ological	morphological
×	replace damaged letter	holo(t)ype	holotype
□/	inset one em	*Mycoblastus* Millstone grit	*Mycoblastus* Millstone grit
▭/	inset two ems	*Mycoblastus* Millstone grit	*Mycoblastus* Millstone grit
run on	run on	is discovered. ⌐ ⌐Fries,	is discovered. Fries,
⌐	move to right	Peziza verrucaria Gliocladium fimbriatum Metarhizium glutinosum	*Peziza verrucaria* *Gliocladium fimbriatum* *Metarhizium glutinosum*
⌐	move to left	*Peziza verrucaria* *Gliocladium fimbriatum* *Metarhizium glutinosum*	*Peziza verrucaria* *Gliocladium fimbriatum* *Metarhizium glutinosum*
⊙	invert type	basidios(d)ores	basidiospores
stet	leave as set (do not make change indicated)	Persoon/indicated/	Persoon indicated
n.p.	start new paragraph	(1821). ⌐Berkeley, however,	(1821). Berkeley, however,

increase the likelihood of an author being in touch with recent work published in his field.

If the reprints do not bear the journal title, volume and page numbers, and correct date of publication the author should add these data before distributing them.

Errors

Any errors in a paper are an embarrassment both to the author and to the editor of a journal. In some cases journals include 'errata slips' in the journals but these are unsatisfactory as they easily become detached and lost. If they are inserted in subsequent issues of the journal the chance of a reader seeing them is only slight. The only reliable way to correct an error and make it widely known is to mention it specifically in a subsequent publication. The correction of errors is an extremely important aspect of all scientific research and all mycologists should endeavour to ensure that any errors in their publications are corrected. From time to time errors will inevitably arise but this should not serve to discourage the student but only to encourage him to try to be more conscientious in future. No mycologist is infallible and it is usually possible to find some small typographical error or other minor slip (e.g. a wrong page number in lists of references) in most systematic publications even by well known specialists.

V. NOMENCLATURE

INTRODUCTION

THE name of an organism is the key to its literature. Consequently workers in all parts of the world should ideally use the same name for each particular plant. In order to achieve any degree of stability in the application of names and conformity in the description of newly discovered taxa it is essential that taxonomists throughout the world adhere to a strict code of practice. A number of different codes exist for different groups of organisms (see Jeffrey, 1973) but the nomenclature of fungi and lichens is controlled by the International Code of Botanical Nomenclature* which is also concerned with vascular plants, bryophytes, algae, all fossil plants, and to some extent the bacteria. This Code stems from a series of rules, the 'Lois de la Nomenclature Botanique', proposed by A. P. De Candolle at the International Botanical Congress held in Paris in 1867. Each subsequent International Botanical Congress considers any amendments and revisions proposed since the last Congress, accepts or rejects them (often on the basis of recommendations by 'Special Committees'), and publishes a new edition of the Code. All botanists are obliged to adhere to the most recent edition of the Code in their work and any who fail to do so are likely to find their work largely ignored or severely criticized and any new names they propose not taken up by other workers.

The Code adopted by the Eleventh International Botanical Congress held in Seattle in August 1969 (Stafleu *et al.*, 1972) contains six fundamental Principles, 75 Articles which must be followed provided that their interpretation is not contrary to one of the Principles, numerous Recommendations which although not obligatory should be adhered to wherever possible, three Appendices, and guides to typification and the citation of botanical literature. The Code is presented in English, French and German but while all are official should there be any minor inconsistencies due to language difficulties the English wording is taken as arbitrarily correct. The application of the Code to particular cases is not always easy and frequently extremely time consuming. It is, however, essential that anyone involved in the naming of fungi and lichens should understand (a) the principles involved; (b) how to set about ascertaining the correct name of a taxon when confronted with a list of names referring to it; and (c) how to make nomenclatural changes, introduce new names, and describe new taxa.

In 1876 several of the leading cryptogamists of that time (Berkeley *et al.*, 1876) joined together to write a brief note criticizing some of their colleagues for not adhering to the recommended nomenclatural practices. A glance at recent issues of the *Index of Fungi*, in which invalidly published and incorrectly formed

* referred to here as the 'Code'.

124

names are indicated, shows that nomenclatural vagrants are still all too plentiful in mycology today.

Newcomers to systematic mycology frequently find that the Code itself is a rather difficult document to interpret. The way in which the Code operates has been interpreted by Jeffrey (1973) as a 'nomenclatural filter' (Fig. 22). The present section considers each of the major steps illustrated in Fig. 22 in turn and explains some of the ways in which they are operated in practice. This section is only intended as a brief introduction to the Code and all possible complications which may arise are not discussed. Parts relevant to particular ranks only have already been reviewed on pp. 38–47.

To obtain a thorough understanding of the Code it is necessary to study its application in the resolving of particular problems. In the following section (pp. 131–178) the portions of the Code relevant to mycologists and lichenologists are reproduced together with examples drawn from their fields. A glossary of terms employed in mycological nomenclature is given on pp. 200–207.

A brief introduction to fungal nomenclature is provided by Ainsworth (1973) and Jeffrey (1973) gives a useful general account of biological nomenclature.

Publication

Only names which have been validly published require consideration under the Code. Handwritten names on herbarium packets which have never appeared in the literature can consequently be ignored; if an author does not accept any such herbarium names and does not wish to introduce them by validating them they should not be mentioned in print (*cf*. Fig. 6). Many names which have appeared in print are, nevertheless, still not validly published. Firstly, all names which are to be considered must be effectively published; i.e. they must have been included in printed matter distributed at least to botanical institutions (Arts. 29, 31). Particular attention should be paid at this stage to the date of effective publication (Art. 30; see also pp. 110–111). Even though a name has been effectively published this does not mean that it is validly published as this depends on the written matter accompanying the name. To be validly published (Art. 32) the name must also have a form which agrees with Arts. 16–24, 26–27, be accompanied by a description or reference to a previously validly published description of the taxon, and comply with the special criteria of Arts. 33–45. Names which are not accepted by the publishing author are not validly published (Art. 34). The rank of the taxon must be made clear in names appearing after 1 January 1953 (Art. 35), and for names appearing after 1 January 1935 the description or diagnosis must be in Latin, accompanied by one in Latin, or refer to a previously validly published description (Art. 36). After 1 January 1958 the nomenclatural type must also be cited for the name to be validly published.

If any single requirement for valid publication is not satisfied the name is not validly published. The name of the author who validly publishes a name is always cited after the name in systematic works, often in an abbreviated form (see pp. 180–187; Arts. 46–49). If the name was previously published invalidly and the validating author attributes it to him the word 'ex' is used to link the names of the two authors (Recs. 46C, E). The name of the validating author is

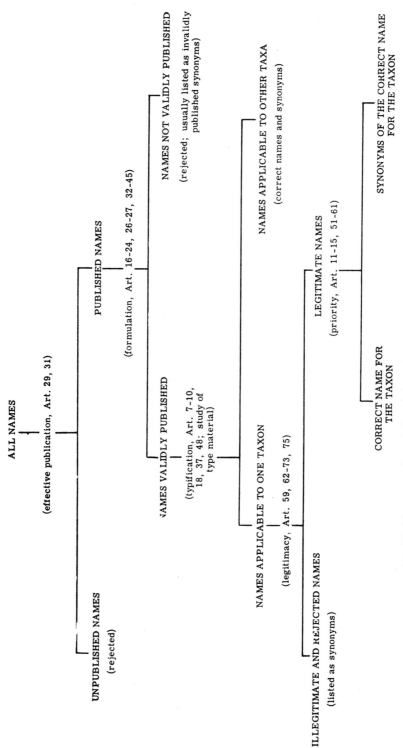

Fig. 22. The nomenclatural filter (adapted and modified from Jeffrey, 1973).

ALL NAMES

(effective publication, Art. 29, 31)

UNPUBLISHED NAMES
(rejected)

PUBLISHED NAMES

(formulation, Art. 16–24, 26–27, 32–45)

NAMES NOT VALIDLY PUBLISHED

(rejected; usually listed as invalidly
published synonyms)

NAMES VALIDLY PUBLISHED

(typification, Art. 7–10,
18, 37, 48; study of
type material)

NAMES APPLICABLE TO ONE TAXON

(legitimacy, Art. 59, 62–73, 75)

NAMES APPLICABLE TO OTHER TAXA

(correct names and synonyms)

LEGITIMATE NAMES

(priority, Art. 11–15, 51–61)

ILLEGITIMATE AND REJECTED NAMES

(listed as synonyms)

CORRECT NAME FOR
THE TAXON

SYNONYMS OF THE CORRECT NAME
FOR THE TAXON

the most important here. When a taxon has its generic position or rank changed the name of the author of the original epithet is placed in brackets and the name of the author making the change after the bracket. To be validly published changes of this type (i.e. new combinations) made after 1 January 1953 must be accompanied by the citation of the name-bringing epithet (i.e. basionym) together with its full bibliographic citation (Art. 33).

Typification

The purpose of typification is to fix permanently the application of names of all ranks governed by the Code so as to preclude the possibility of the same name being used in different senses; i.e. for different plants. Typification is consequently the keystone of stability in the application of names. The type of a family name is a genus (Arts. 10, 18), that of a genus a species (Art. 10) and that of a species or infraspecific taxon a dried herbarium specimen (Art. 9).

As has been mentioned in the preceding section, to be validly published new names now have to be accompanied by the designation of a nomenclatural type (Art. 37). A specimen designated by the describing author as the type for his name is called the *holotype* (Art. 7). A common error in mycological publications is to designate living cultures as the holotypes for species but this is not permissible under the Code (Art. 9); dried cultures permanently preserved in herbaria are, however, acceptable as types. Living cultures derived from the plate or specimen dried as the holotype and maintained in culture collections should be referred to as 'ex-holotype' (or 'culto-type' as suggested by van Beverwijk, 1961). 'Type cultures', in the sense of the use of this term by bacteriologists, do not exist for the fungi. The series of terms proposed by Ciferri (1957) to designate kinds of 'type cultures' has consequently not been followed by later mycologists. Holotypes should always be preserved in a major mycological herbarium from which they may be borrowed by other workers who might wish to examine them and authors describing new taxa are recommended to indicate where the material has been deposited (Rec. 7B). Where sufficient material is available it is a good policy to distribute parts of it to the principal mycological herbaria; such duplicates of the holotype collection are termed *isotypes*. While isotypes do not have the same nomenclatural importance as holotypes they may assume considerable importance if the holotype is lost or destroyed.

When a newly described species is known from several collections in addition to the holotype these will also usually be cited by the describing author. Such collections are referred to as *paratypes* and, like isotypes, should be carefully preserved in case the holotype (and isotypes if any were extant) is lost or destroyed.

Authors publishing new names before 1 January 1958 were not required to designate nomenclatural types (Art. 37) and frequently did not. If only a single taxon or specimen was referred to in the original place of publication this may in most instances be taken as a holotype. Often, however, not one but several species or specimens were listed in the original place of publication; these are termed *syntypes*. Later authors working on the group, after a careful study of the published description, any discussion, and illustrations and plates (both of the describing author and any he might have referred to), and the application of the name by other authors, are obliged to select one of the syntypes to serve as a type, termed a *lectotype*. In the case of species and infraspecific taxa in the absence of any specimens in the author's herbarium in some cases illustrations either provided by the original author or referred to by him can be designated as lecto-types (Art. 9 Note 1). Early post-starting point authors frequently refer to pre-starting point bi- or polynomials and illustrations which can lead to material suitable for lectotypification (e.g. Linnaeus, 1753, referring to Dillenius, 1741; Fries, 1821, referring to Acharius, 1810; see e.g. Hawksworth and Punithalingam, 1973).

Authors wishing to typify untypified names must first of all do everything they can to trace any material either used by the describing author in compiling his description or named by him. To do this some knowledge of the location of herbaria is required (see pp. 179–192). If an extensive study shows that the holotype (if there ever was one) has been lost or destroyed and there is no suitable material studied by the describing author before he published the taxon suitable for lectotypification available, a *neotype* must be designated. A 'neotype' is simply a specimen or other element which serves as the nomenclatural type in the absence of a holotype or lectotype. The selection of a neotype requires very care-ful consideration and should only be carried out by someone very familiar with the group concerned after an exacting study of the original description, data referred to in it, and the use of the name by later authors. The same criteria employed in designating a holotype should be adopted when selecting a neotype (see p. 65). In some cases exsiccatae already issued may be used in neotypifications so ensuring that duplicates (i.e. 'isoneotypes') are present in the major herbaria.

Once a lectotype or neotype has been designated for a name this choice must be followed unless it can be shown that the choice was based on a misunder-standing of the original description (Art. 8). If material suitable for lectotypifica-tion comes to light subsequent to the designation of a neotype, a lectotype, which has priority over a neotype, must be chosen from it. Similarly, if a holotype is discovered after the selection of a lectotype or neotype this assumes priority as the nomenclatural type of the taxon.

In some cases the original material on which a taxon was based proves to be heterogeneous (i.e. it includes more than one taxon according to modern concepts). Where one part of the material was clearly intended as the type this presents no great problem but where none of it agrees with the original descrip-tion or this was based on both elements the situation may be more complex. If neither element can be selected as a satisfactory type the name must be rejected (Art. 70).

In typifying names the provisions of the Code should be carefully followed and all names judged on their own merits. In some cases lectotypifications have been carried out primarily to dispose of unwanted names. This constitutes an abuse of the type method and must be condemned, as when later workers discover what has occurred and rectify the situation this may lead to further name changes.

If the type of a name is excluded from an account of that taxon (e.g. the type species of a genus is removed from that genus and placed in another although the genus of which it is the type is also accepted) the author making this change must be regarding the name as based on another type (even if none was specifically designated) and so has inadvertently created a homonym (see p. 161; Art. 48). If, when describing a new genus or species, the describing author cites a species or specimen, respectively, which is already the type of a validly published legitimate name he is considered as having published a superfluous name (Art. 63) and his name must be typified by the type of the name he ought to have adopted (Art. 7).

For further information on the typification of names see the 'Guide for the determination of types' in the Code (Stafleu *et al.*, 1972).

Legitimacy

When all validly published and correctly formulated names have been typified the next stage (Fig. 22) is to determine which are legitimate (i.e. names which must be taken into account in determining the correct name for a taxon) and which are illegitimate (i.e. names which must not be taken into consideration but must be treated as synonyms).

A name is illegitimate and must be rejected if it was nomenclaturally superfluous when published (see p. 168; Art. 63), if it is spelt exactly like a previously validly published taxon of the same rank (i.e. a homonym) in any group of plants (Art. 64), if it has been used in different senses and become a persistent source of error (Art. 69), if it is based on discordant elements (Art. 70), if it is based on a monstrosity (Art. 71), or if it contravenes any of Arts. 51, 53–58 and 60 (Arts. 66, 67). In the case of fungi with two or more stages in their life cycles if a name typified by an imperfect state specimen is employed in a perfect state genus (i.e. a genus whose type species is typified by a perfect state specimen) this name is treated as illegitimate (Art. 59).

In some cases an illegitimate epithet which was validly published may become legitimate on transference to a new rank or position but it is then attributed to the author introducing the legitimate name without reference to the author who first used it illegitimately (Art. 72). Specific epithets are not themselves illegitimate if published under an illegitimate generic name (Art. 68).

Newcomers to nomenclature often find some of the reasons why names are regarded as illegitimate and rejected difficult to understand and the examples given in the following section under the Articles referred to above should be carefully studied.

Priority

The nomenclature of all taxonomic groups covered in the Code is based primarily on the priority of publication of all validly published legitimate names in the same rank which apply to the particular taxon under consideration (Art. 11). This principle of priority is, however, limited in two main ways. Firstly, by means of starting point dates for particular groups (Art. 13), all names which appeared before the starting points being treated as not validly published (although the same names may be validated by works after the appropriate starting point date). The appropriate date for the starting point of any particular name is determined by the group to which its nomenclatural type belongs and not by the group to which the publishing author assigned it where these differ (Art. 13 Note 1). Secondly, in the case of fungi with pleomorphic life cycles, perfect and imperfect state names have to be considered independently from one another for nomenclatural purposes; i.e. the correct name for a fungus with one or more states is the earliest legitimate name typified by the perfect state but imperfect state names may be used in works dealing with those states (Art. 59).

If an earlier name for a well known genus is found conservation of the latter can be proposed (see below) but for ranks below that of genus the earliest legitimate validly published epithet must be employed. If the epithet is not combined with the appropriate generic name a new combination will be required.

All other available (i.e. legitimate and validly published) names are then treated as synonyms of the correct name. Synonyms themselves fall into two categories which are often distinguished in published lists of synonyms (see p. 66; Fig. 7). 'Nomenclatural synonyms' ('obligate synonyms', 'typonyms', 'synisonyms' or 'homotypic synonyms') are synonyms based on the same nomenclatural type so that they are synonyms regardless of taxonomic judgement. 'Taxonomic synonyms' ('facultative synonyms', 'heterotypic synonyms'), in contrast, are based on different nomenclatural types and consequently it is the taxonomists judgement which determines to which taxon they apply. All epithets in the same rank as that of the correct name for the taxon which are available but of a more recent date are sometimes referred to as 'younger synonyms'.

Conservation

Names in the rank of family and genus can be conserved against other names which would have to be adopted if the Code were strictly applied to them, to prevent disadvantageous changes in nomenclature (Art. 14). The importance of conserving well established names has already been referred to (p. 41). Lists of conserved names are published as an Appendix to the Code and additions may be proposed. The current Code lists 58 conserved fungal and 32 conserved lichen generic names. Each proposal for conservation is considered by the Special Committee for Fungi and Lichens, the members of which vote on the proposal and inform the General Committee of their opinion. If their recommendation is accepted by the General Committee botanists are authorized to continue to use the name pending the decision of the next International Botanical Congress which will be asked to ratify their recommendation (Art. 15).

In some cases when the type material of the type species of a well known genus is examined it is found to belong to a genus other than that of which it has been taken to be a member. In such cases it is possible to invoke the conservation procedures to conserve a generic name with a type other than that which should be adopted for it by the strict application of the Code (Art. 14). In some cases such a conserved name may have to be attributed to an author other than the one who originally described it (e.g. the first author to use the name in its accepted sense and exclude the type of the original name).

THE INTERNATIONAL CODE OF BOTANICAL NOMENCLATURE

IN this section the parts of the International Code of Botanical Nomenclature adopted by the Eleventh International Botanical Congress held at Seattle in August 1969 (Stafleu *et al.*, 1972) pertinent to mycologists and lichenologists are reproduced. The interpretation of the various Articles and Recommendations has been illustrated by examples drawn from these fields and some additional explanatory notes have also been included. All information included within square brackets has been introduced here for the benefit of mycologists and lichenologists entering the field of taxonomic mycology and does not constitute part of the Code itself (although most mycological examples from the Code are retained).

Anyone involved in taxonomic revisions or who becomes confronted by nomenclatural problems is recommended to study these examples carefully (preferably with the original publications mentioned to hand), and also to consult the complete version of the Code together with its authorized examples drawn mainly from vascular plant situations.

INTERNATIONAL CODE OF BOTANICAL NOMENCLATURE
PREAMBLE

Botany requires a precise and simple system of nomenclature used by botanists in all countries, dealing on the one hand with the terms which denote the ranks of taxonomic groups or units, and on the other hand with the scientific names which are applied to the individual taxonomic groups of plants. The purpose of giving a name to a taxonomic group is not to indicate its characters or history, but to supply a means of referring to it and to indicate its taxonomic rank. This Code aims at the provision of a stable method of naming taxonomic groups, avoiding and rejecting the use of names which may cause error or ambiguity or throw science into confusion. Next in importance is the avoidance of the useless creation of names. Other considerations, such as absolute grammatical correctness, regularity or euphony of names, more or less prevailing custom, regard for persons, etc., notwithstanding their undeniable importance, are relatively accessory.

The *Principles* form the basis of the system of botanical nomenclature.

The detailed provisions are divided into *Rules*, set out in the Articles, and *Recommendations*; the notes attached to these are integral parts of them. Examples are added to the rules and recommendations to illustrate them.

The object of the *Rules* is to put the nomenclature of the past into order and to provide for that of the future; names contrary to a rule cannot be maintained.

The *Recommendations* deal with subsidiary points, their object being to bring about greater uniformity and clearness, especially in future nomenclature; names contrary to a recommendation cannot, on that account, be rejected, but they are not examples to be followed.

The provisions regulating the modification of this Code form its last division.

The Rules and Recommendations apply throughout the plant kingdom, recent and fossil. However, special provisions are needed for certain groups.

[NOTE: References to other Codes included here.]

The only proper reasons for changing a name are either a more profound knowledge of the facts resulting from adequate taxonomic study or the necessity of giving up a nomenclature that is contrary to the rules.

In the absence of a relevant rule or where the consequences of rules are doubtful, established custom is followed.

This edition of the Code supersedes all previous editions.

Division I. Principles
Principle I
Botanical nomenclature is independent of zoological nomenclature.

The Code applies equally to names of taxonomic groups treated as plants whether or not these groups were originally assigned to the plant kingdom.

[NOTE: For a comparison of existing accepted Codes of nomenclature for different groups or organisms see Jeffrey (1973).]

Principle II
The application of names of taxonomic groups is determined by means of nomenclatural types.

Principle III
The nomenclature of a taxonomic group is based upon priority of publication.

Principle IV
Each taxonomic group with a particular circumscription, position, and rank can bear only one correct name, the earliest that is in accordance with the Rules, except in specified cases.

Principle V
Scientific names of taxonomic groups are treated as Latin regardless of their derivation.

Principle VI
The Rules of nomenclature are retroactive unless expressly limited.

[I.E.: They affect all names even if published before a particular Article came into being unless specific dates are listed in particular Articles.]

beringiana, Cladonia submitis f. *divaricata, Stereocaulon vesuvianum* var. *nodulosum* f. *sessile, Parmelia* subgen. *Amphigymnia* sect. *Subflavescentes* ser. *Emaculatae.*]

Section 2. TYPIFICATION
Article 7

The application of names of taxa of the rank of family or below is determined by means of *nomenclatural types* (types of names of taxa). A nomenclatural type (*typus*) is that constituent element of a taxon to which the name of the taxon is permanently attached, whether as a correct name or as a synonym.

Note 1. The nomenclatural type is not necessarily the most typical or representative element of a taxon; it is that element with which the name is permanently associated.

A *holotype* is the one specimen or other element used by the author or designated by him as the nomenclatural type. As long as a holotype is extant, it automatically fixes the application of the name concerned.

If no holotype was indicated by the author who described a taxon, or when the holotype has been lost or destroyed, a *lectotype* or a *neotype* as a substitute for it may be designated. A lectotype always takes precedence over a neotype. An *isotype*, if such exists, must be chosen as the lectotype. If no isotype exists, the lectotype must be chosen from among the *syntypes*, if such exist. If neither an isotype nor a syntype nor any of the original material is extant, a neotype may be selected.

A *lectotype* is a specimen or other element selected from the original material to serve as a nomenclatural type when no holotype was designated at the time of publication or as long as it is missing. When two or more specimens have been designated as types by the author of a specific or infra-specific name (e.g. male and female, flowering and fruiting, etc.), the lectotype must be chosen from among them.

An *isotype* is any duplicate (part of a single gathering made by a collector at one time) of the holotype; it is always a specimen.

A *syntype* is any one of two or more specimens cited by the author when no holotype was designated, or any one of two or more specimens simultaneously designated as types.

A *neotype* is a specimen or other element selected to serve as nomenclatural type as long as all of the material on which the name of the taxon was based is missing.

A new name or epithet published as an avowed substitute (*nomen novum*) for an older name or epithet is typified by the type of the older name.

A new name formed from a previously published legitimate name or epithet (*stat. nov., comb. nov.*) is, in all circumstances, typified by the type of the basionym.

A name or epithet which was nomenclaturally superfluous when published (see Art. 63) is automatically typified by the type of the name or epithet which ought to have been adopted under the rules, unless the author of the superfluous name or epithet has indicated a definite type.

The type of a name of a taxon assigned to a group with a nomen-clatural starting-point later than 1753 (see Art. 13) is to be determined in accordance with the indication or description and other matter accompanying its first valid publication (see Arts. 32–45). When valid publication is by reference to a pre-starting-point description, the latter must be used for purposes of typification as though newly published.

A change of the listed type-species of a conserved generic name (see Art. 14 and App. III) can be effected only by a procedure similar to that adopted for the conservation of generic names.

The type of the name of a taxon of fossil plants of the rank of species or below is the specimen whose figure accompanies or is cited in the valid publication of the name (see Art. 38). If figures of more than one specimen were given or cited when the name was validly published, one of those specimens must be chosen as type.

Note 2. The typification of names of genera based on plant mega-fossils and plant microfossils (form- and organ-genera), genera of imperfect fungi, and any other analogous genera or lower taxa does not differ from that indicated above.

[NOTE: See pp. 127–129 for further information on types and typification.]

Recommendation 7A

It is strongly recommended that the material on which the name of a taxon is based, especially the holotype, be deposited in a permanent, responsible institution and that it be scrupulously conserved. When living material is designated as a nomenclatural type (for Bacteria only; see Art. 9, paragraph 3), appropriate parts of it should be immediately preserved.

Recommendation 7B

Whenever the elements on which the name of a taxon is based are hetero-geneous, the lectotype should be so selected as to preserve current usage unless the element thus selected is discordant with the protologue.*

[EXAMPLE: *Nemania* Gray (*Nat. Arr. Br. Pl.* **1**: 516, 1821) comprised 21 species now disposed in *Epichlöe* (Fr.) Tul., *Hypoxylon* Bull. ex Fr., *Nummularia* Tul. and *Ustulina* Tul. House (*Bull. N.Y. St. Mus.* **266**: 48, 1925) selected the first species treated by Gray, *N. ustulatum* Bull. ex Gray (syn. *Ustulina deusta* (Hoffm. ex Fr.) Lind) as lectotype but this means that the well established name *Ustulina* Tul. (*Sel. Carp. Fung.* **2**: 22, 1863) based on this species must be replaced by *Nemania*. As Gray's original description does not accord more closely with the *Ustulina* than several other species treated, Donk (*Regnum Vegetabile* **34**: 16–21, 1964) considered that it should be typified by another of the original species, *N. serpens* (Pers. ex Fr.) Gray (syn. *Hypoxylon serpens* (Pers. ex Fr.)Kickx) so as to preserve current usage.]

Article 8

The author who first designates a lectotype or a neotype must be followed, but his choice is superseded if the holotype or, in the case of a neotype, any of the original material is rediscovered; it may also be

---

* Protologue (from προτος, first λογος, discourse): everything associated with a name at its first publication, i.e. diagnosis, description, illustrations, references, synonymy, geographical data, citation of specimens, discussion, and comments.

superseded if it can be shown that the choice was based upon a misinterpretation of the protologue, or was made arbitrarily.

[EXAMPLE: Earle (*Bull. N. Y. Bot. Gdn* **5**: 383, 447, 1909) selected *Agaricus ramentaceus* Bull. ex Fr. as lectotype for *Armillaria* (Fr. ex Fr.) Staude but this is not acceptable as it was not included in *Agaricus* trib. *Armillaria* Fr. ex Fr. *Agaricus melleus* Vahl ex Fr. was selected by Clements and Shear (*Gen. Fungi*: 348, 1932) and was one of Fries' original species. Singer (*Annls mycol.* **34**: 331, 1936), however, considered *Agaricus luteovirens* Alb. & Schw. ex Fr. to be the type but this choice must be rejected as this was not one of Fries' original species. *Agaricus robustus* Alb. & Schw. ex Fr. was proposed by Imai (*J. Fac. Agr. Hokkaido Univ.* **43**: 47, 1938) but must also be rejected for although one of Fries' original species it does not accord more closely with the original description of *Agaricus* trib. *Armillaria* than the previously selected lectotype, *A. melleus* (see *Beih. Nova Hedwigia* **5**: 31–34, 1962).

'Arbitrarily' as used in Art. 8 means the employment of some mechanical system in selecting the lectotype of a name; e.g. choosing the first species or specimen mentioned without reference to the protologue.]

Article 9

The type (*holotype*, *lectotype*, or *neotype*) of a name of a species or infraspecific taxon is a single specimen or other element except in the following case: for small herbaceous plants and for most non-vascular plants, the type may consist of more than one individual, which ought to be conserved permanently on one herbarium sheet or in one preparation.

If it is later proved that such a type herbarium sheet or preparation contains parts belonging to more than one taxon, the name must remain attached to that part (*lectotype*) which corresponds most nearly with the original description.

[EXAMPLE: The lectotype sheet of *Lichen jubatus* L. (*Sp. Pl.* **2**: 1155, 1753) in LINN includes specimens of two species; one black and one yellowish green. Linnaeus did not specify the colour of his plant but two of the polynomials he cites refer to the species as being 'nigricans'. The dark specimen consequently corresponds most nearly to the protologue and is designated as the lectotype (see *Taxon* **19**: 237, 1970).]

Type specimens of names of taxa, the Bacteria excepted, must be preserved permanently and cannot be living plants or cultures.

[I.E.: Living cultures of fungi cannot be nomenclatural types although dried (dead) cultures preserved in a herbarium may be (see p. 127).]

Note 1. If it is impossible to preserve a specimen as the type of a name of a species or infraspecific taxon of recent* plants, or if such a name is without a type specimen, the type may be a description or figure.

[EXAMPLE: *Septotrullula bacilligera* Höhn. was lectotypified by an illustration, on a herbarium packet, drawn by Höhnel, although the material in the packet contained no determinable fungus, as Höhnel's drawing and description left no doubt as to its identity (see *Trans. Br. mycol. Soc.* **48**: 349–366, 1965).]

Note 2. One whole specimen used in establishing a taxon of fossil plants is to be considered the nomenclatural type. If this specimen is cut into pieces (sections of fossil wood, pieces of coalball plants, etc.),

---

* The term *recent* as used here and elsewhere in the Code is in contradistinction to *fossil* (see Art. 13, Note 2).

all parts originally used in establishing the diagnosis ought to be clearly
marked.

Article 10

The type of a name of a genus or of any taxon between genus and
species is a species, that of a name of a family or of any taxon between
family and genus is the genus on whose present or former name that of
the taxon concerned is based (see also Art. 18).

The principle of typification does not apply to names of taxa above
the rank of family (see Art. 16).

Note 1. The type of a name of a family not based on a generic name
is the genus that typifies the alternative name of that family (see Art. 18).

Note 2. For the typification of some names of subdivisions of
genera* see Art. 22.

Section 3. PRIORITY

Article 11

Each family or taxon of lower rank with a particular circumscription,
position, and rank can bear only one correct name, special exceptions
being made for 9 families for which alternative names are permitted
[vascular plants], and for certain fungi and fossil plants (see Art. 59).

For any taxon from family to genus inclusive, the correct name is the
earliest legitimate one with the same rank, except in cases of limitation
of priority by conservation (see Arts. 14 and 15) or where Arts. 13f, 58,
or 59 apply.

For any taxon below the rank of genus, the correct name is the
combination of the earliest available legitimate epithet in the same rank
with the correct name of the genus or species to which it is assigned,
except where Arts. 13f, 22, 26, 58, or 59 apply.

The principle of priority does not apply to names of taxa above the
rank of family (see Art. 16).

[NOTE: Nomenclature cannot specify the genus under which a taxonomist should
place a species as this is a matter of opinion. A single species can consequently have
nomenclaturally correct names under more than one generic name.]

Article 12

A name of a taxon has no status under this Code unless it is validly
published (see Arts. 32–45).

[NOTE: A name which has not been validly published does not have to be taken into
consideration for nomenclatural purposes (see Fig. 22) but the same name should be
avoided for a new species although the new species would not be a later homonym
under Art. 64 as the earlier name was not validly published.]

Section 4. LIMITATION OF THE PRINCIPLE OF PRIORITY

Article 13

Valid publication of names for plants of the different groups is

---

* Here and elsewhere in the Code the phrase 'subdivision of a genus' refers only to
  taxa between genus and species in rank.

treated as beginning at the following dates (for each group a work is mentioned which is treated as having been published on the date given for that group):

Recent plants

d. LICHENES, 1 May 1753 (Linnaeus, *Species Plantarum* ed. 1). For nomenclatural purposes names given to lichens shall be considered as applying to their fungal components.*

e. FUNGI: UREDINALES, USTILAGINALES, and GASTEROMYCETES, 31 Dec. 1801 (Persoon, *Synopsis Methodica Fungorum*).

f. FUNGI CAETERI, 1 Jan. 1821 (Fries, *Systema Mycologicum* vol. 1). Vol. 1 of the *Systema* is treated as having appeared on 1 Jan. 1821, and the *Elenchus Fungorum* (1828) is treated as a part of the *Systema*. Names of FUNGI CAETERI published in other works between the dates of the first (vol. 1) and last (vol. 3, part 2 and index) parts of the *Systema* which are synonyms or homonyms of names of any of the FUNGI CAETERI included in the *Systema* do not affect the nomenclatural status of names used by Fries in this work.

h. MYXOMYCETES, 1 May 1753 (Linnaeus, *Species Plantarum* ed. 1).

Fossil plants

j. ALL GROUPS, 31 Dec. 1820 (Sternberg, *Flora der Vorwelt, Versuch* 1: 1–24. *t. 1–13*). Schlotheim, *Petrefactenkunde*, 1820, is regarded as published before 31 Dec. 1820.

Note 1. The group to which a name is assigned for the purposes of this Article is determined by the accepted taxonomic position of the type of the name.

[EXAMPLES: *Lichen rugosum* L. (*Sp. Pl.* **2**: 1140, 1753) is treated as not validly published as although it was included by Linnaeus as a lichen it is typified by a non-lichenized fungus (see *Trans. Br. mycol. Soc.* **60**: 501–509, 1973); *Mucor furfuraceus* L. (*Sp. Pl.* **2**: 1185, 1753), in contrast, was included by Linnaeus amongst the Fungi but this epithet was validly published as it is typified by a lichenized and not a non-lichenized fungus.]

Note 2. Whether a name applies to a taxon of fossil plants or of recent plants is decided by reference to the specimen that serves directly or indirectly as its nomenclatural type. The name of a species or infraspecific taxon is treated as pertaining to a recent taxon unless its type specimen is fossil in origin. Fossil material is distinguished from recent material by stratigraphic relations at the site of original

---

* [NOTE: From this it follows that lichen algae (phycobionts) can be named independently from the lichen they form a part of but that the fungal components cannot be given separate new names as that of the lichen is considered to refer to the fungal partner (mycobiont). The numerous new generic names ending in the suffix '*-myces*' proposed by Thomas (*Beitr. Krypt.-Fl. schweiz* **9**(1), 1939) and Ciferri and Tomaselli (*Atti Ist. Bot. Lab. Critt. Pavia, ser.* 5, **10**, 1953) for the fungal components of lichens to replace the names employed for the intact thalli (e.g. *Ramalinomyces* Thom. ex Cif. & Tom. for the fungal component of *Ramalina* Ach.) are consequently superfluous names which must be rejected under Art. 63.]

occurrence. In cases of doubtful stratigraphic relations, regulations for recent taxa shall apply.

Note 4. The two volumes of Linnaeus' *Species Plantarum* ed. 1 (1753), which appeared in May and August, 1753, respectively, are treated as having been published simultaneously on the former date (1 May 1753).

[NOTE: This Article is extremely important for mycologists because of the different starting point dates for nomenclature in different groups. Precise dates of publication within works appearing in 1821 particularly are often needed to ascertain who validated a particular pre-starting point name first (see *Taxon* **6**: 245–256, 1957). It should be noted that pre-starting point date works reissued after those dates which are unaltered are not to be accepted as valid; a 'second edition' of Fries' *Observationes mycologicae*, originally published in 1815–18, for example, was reissued in 1824 with only a changed title page and is not considered to validate the names in the original issue (see *Mycologia* **31**: 297–307, 1939). 31 Dec. 1801 was proposed as the starting point date for Hypho-mycetes (see *Taxon* **8**: 96–103, 1959) but this proposal was not accepted. Pre-starting point names which meet the other requirements of the Code for valid publication are often referred to as 'devalidated names' (see p. 201). EXAMPLES: *Agaricus* L. (*Sp. Plant.* **2**: 1171, 1753) was not validly published as it appeared prior to 1 Jan. 1821 and this name was first validated by Fries (*Syst. mycol.* **1**: lvi, 8, 1821); *Sphaeria deusta* Hoffm. (*Veg. crypt.* **1**: 3, 1787) was validated by Fries (*Syst. mycol.* **2**: 345, 1823). For the author citation of pre-starting point subsequently validated names see under Rec. 46E.

The phrase '. . . the nomenclatural status of names used by Fries in this work' (Art. 13f) means that the names used by Fries must be taken up in preference to those not accepted or not in the starting-point works even if of an earlier date.]

Article 14

In order to avoid disadvantageous changes in the nomenclature of genera, families, and intermediate taxa entailed by the strict applica-tion of the rules, and especially of the principle of priority in starting from the dates given in Art. 13, this Code provides, in Appendices II and III, lists of names that are conserved (*nomina conservanda*) and must be retained as useful exceptions. Conservation aims at retention of those generic names which best serve stability of nomenclature. (See Rec. 50E.)

Note 1. These lists of conserved names will remain permanently open for additions. Any proposal of an additional name must be accom-panied by a detailed statement of the cases both for and against its conservation. Such proposals must be submitted to the General Committee (see Division III), which will refer them for examination to the committees for the various taxonomic groups.

Note 2. The application of both conserved and rejected names is determined by nomenclatural types.

Note 3. A conserved name is conserved against all other names in the same rank based on the same type (nomenclatural synonyms, which are to be rejected) whether these are cited in the corresponding list of rejected names or not, and against those names based on different types (taxonomic synonyms) that are cited in that list. When a conserved name competes with one or more other names based on different types and against which it is not explicitly conserved, the earliest of the competing names is adopted in accordance with Art. 57.

[EXAMPLE: If the genus *Macropyrenium* Hampe ex Massal. (*Atti Ist. Veneto, ser.* 3, **5**: 329, 1860) is united with *Phaeotrema* Müll. Arg. (*Mém. Soc. Phys. hist. nat. Genève* **29**(8): 10, 1887) the combined genus should bear the earlier generic name *Macropyrenium* although *Phaeotrema* is a conserved name (against *Asteristion* Leight., *Trans. Linn. Soc. Lond.* **27**: 163, 1870) and *Macropyrenium* is not. In such cases, in the interests of nomenclatural stability, conservation against the earlier unlisted name should be sought unless the earlier name is already well established.]

Note 4. When a name of a genus has been conserved against an earlier name based on a different type, the latter is to be restored, subject to Art. 11, if it is considered the name of a genus distinct from that of the *nomen conservandum* except when the earlier rejected name is a homonym of the conserved name.

[EXAMPLE: The generic name *Pleurotus* (Fr.) Kumm. (*Führ. Pilzk.* 24, 104, 1871) is conserved against four earlier names including *Pterophyllus* Lév. (*Ann. Sci. Nat., Bot., sér.* 3, **2**: 178, 1844) which is based on a different nomenclatural type. If *Pterophyllus* Lév. were considered to be a genus distinct from *Pleurotus* (Fr.) Kumm., the name *Pterophyllus* Lév. could be retained for it.]

Note 5. A conserved name is conserved against all its earlier homonyms.

[EXAMPLE: If a validly published generic name *Podospora* described prior to 1856 were discovered the conserved generic name *Podospora* Ces. (in Rabenh., *Klotzsch. Herb. Mycol.*, ed. 2, no. 258, 1856) would not be affected as it is considered to be conserved against any earlier homonyms.]

Note 6. Provision for the conservation of a name in a sense that excludes the original type is made in Art. 48.

[NOTE: See example cited under Art. 48 (p. 161).]

Note 7. When a name is conserved only to preserve a particular orthography, it is to be attributed without change of priority to the author who originally described the taxon.

Article 15

When a name proposed for conservation has been approved by the General Committee after study by the Committee for the taxonomic group concerned, botanists are authorized to retain it pending the decision of a later International Botanical Congress.

Recommendation 15A

When a name proposed for conservation has been referred to the appropriate Committee for study, botanists should follow existing usage as far as possible pending the General Committee's recommendation on the proposal.

[EXAMPLE: The generic name *Volutella* Fr. (*Syst. mycol.* **3**: 466, 1832) has been proposed for conservation against the earlier homonym *Volutella* Forsk. (*Fl. Aeg.-Arab.*: 84, 1775) (see *Taxon* **21**: 707–708, 1972). Mycologists are recommended to continue to use *Volutella* Fr. pending the General Committee's recommendation. Such names proposed for conservation are usually cited in the form '*Volutella* Fr. *nom. cons. prop.*' pending the decision of a subsequent International Botanical Congress.]

Chapter III. NOMENCLATURE OF TAXA ACCORDING TO THEIR RANK

Section 1. NAMES OF TAXA ABOVE THE RANK OF FAMILY

Article 16

The principles of priority and typification do not affect the form of names of taxa above the rank of family. (See Arts. 10 and 11.)

Recommendation 16A

(a) The name of a division is preferably taken from characters indicating the nature of the division as closely as possible; it should end in *-phyta*, except when it is a division of FUNGI, in which case it should end in *-mycota*. Words of Greek origin are generally preferable.

The name of a subdivision is formed in a similar manner; it is distinguished from a divisional name by an appropriate prefix or suffix or by the ending *-phytina*, except when it is a subdivision of FUNGI, in which case it should end in *-mycotina*.

(b) The name of a class or of a subclass is formed in a similar manner and should end as follows:

2. In the FUNGI: *-mycetes* (class) and *-mycetidae* (subclass).

Article 17

If the name of an order is based on the stem of a name of a family, it must have the ending *-ales*. If the name of a suborder is based on the stem of a name of a family, it must have the ending *-ineae*.

Note 1. Names intended as names of orders, but published with their rank denoted by a term such as 'Cohors', 'Nixus', 'Alliance', or 'Reihe' instead of *ordo* are treated as having been published as names of orders.

Note 2. When the name of an order or suborder based on the stem of a name of a family has been published with an improper termination, the ending is to be changed to accord with the rule, without change of the author's name.

[EXAMPLES of names of orders are Agaricales, Sphaeriales and Ustilaginales (see also p. 39).]

Recommendation 17A

Authors should not published new names of orders for taxa of that rank which include a family from whose name an existing ordinal name is derived.

Section 2. NAMES OF FAMILIES AND SUBFAMILIES, TRIBES AND SUBTRIBES

Article 18

The name of a family is a plural adjective used as a substantive; it is formed by adding the suffix *-aceae* to the stem of a legitimate name of an included genus (see also Art. 10). (For the treatment of final vowels of stems in composition, see Rec. 73G).

[EXAMPLES: Capnodiaceae (Sacc.) Höhn. based on *Capnodium* Mont., Caloplacaceae Zahlbr. based on *Caloplaca* Th. Fr., Hypodermataceae Rehm based on *Hypoderma* DC. em. De Not., Lecanactidaceae Stiz. based on *Lecanactis* Eschw. are correctly formed. Euphacidiaceae Rehm based on *Phacidium* Fr., and Thyreomycetaceae Bonord. based on *Leptostroma* Fr., *Leptothyrium* Kunze ex Wallr.. *Actinothyrium* Kunze and

Microthyrium Desm., however, were formed contrary to this Article and have to be rejected under Art. 32. For further examples see Cooke and Hawksworth (1970).]

Names intended as names of families, but published with their rank denoted by one of the terms order (*ordo*) or natural order (*ordo naturalis*) instead of family, are treated as having been published as names of families.

[EXAMPLE: The 'ordre Lecideae Chev.' (*Fl. env. Paris*: 549, 1826) was intended as a family name although referred to as an 'ordre' and so must be accepted as a family name with its spelling corrected to 'Lecideaceae Chev.' in accordance with Note 2 below.]

Note 1. A name of a family based on the stem of an illegitimate generic name is illegitimate unless conserved. Contrary to Art. 32 (2) such a name is validly published if it complies with the other requirements for valid publication.

[EXAMPLE: The generic name *Peltophora* Clem. is illegitimate under Art. 62 and 63. As the family name Peltophoraceae Clem. has not been conserved it is also regarded as illegitimate.]

Note 2. When a name of a family has been published with an improper termination, the ending is to be changed to accord with the rule, without change of the author's name. (See Art. 32, Note 1.)

[EXAMPLES: The family names 'Endomycetacei Schröt.' and 'Geoglosseae Corda' must have their endings corrected and be cited as the Endomycetaceae Schröt. and Geoglossaceae Corda, respectively. See also Cooke and Hawksworth (1970).]

Article 19

The name of a subfamily is a plural adjective used as a substantive; it is formed by adding the suffix -*oideae* to the stem of a legitimate name of an included genus.

A tribe is designated in a similar manner, with the ending -*eae*, and a subtribe similarly with the ending -*inae*.

The name of any taxon of a rank below family and above genus which includes the type genus of the correct name of the family to which it is assigned is to be based on the name of that genus, but without the citation of an author's name (see Art. 46). This provision applies only to the names of those taxa which include the type of the correct name of the family; the type of the correct name of each such taxon is the same as that of the correct name of the family.

Note 1. Names of other taxa of a rank below family and above genus are subject to the provisions of priority.

[EXAMPLE: The subfamily including the type of the family Ascobolaceae Sacc. (i.e. the genus *Ascobolus* Pers. ex Hook.) is called subfamily Ascoboloideae, while the subfamily which includes *Thelebolus* Tode ex Fr. (and not *Ascobolus*) has to be called subfamily Theleboideae Brumm.]

The first valid publication of a name of a taxon at a rank below family and above genus which does not include the type of the correct name of the family automatically establishes the name of another taxon at the same rank which does include that type. Such autonyms (automatically established names) are not to be taken into consideration for purposes of priority. However, when no earlier name is available, they may be adopted as new in another position.

The name of a subdivision of a family may not be based on the same stem of a generic name as is the name of the family or of any subdivision of the same family unless it has the same type as that name.

[EXAMPLE: The valid publication of Coronophoraceae Höhn. subfamily Nitschkioideae Fitzp. (*Mycologia* **15**: 26, 1923) automatically established the subfamily Coronophoroideae typified by the type of the family name Coronophoraceae (i.e. *Coronophora* Fuckel).]

Note 2. When a name of a taxon assigned to one of the above categories has been published with an improper termination, such as -*eae* for a subfamily or -*oideae* for a tribe, the ending must be changed to accord with the rule, without change of the author's name. However, when the rank of the group is changed by a later author, his name is then cited as author for the name with the appropriate ending, in the usual way.

[EXAMPLE: The subfamily name 'Nitschkieae Fitzp.' (*Mycologia* **15**: 26, 1923) must have its ending corrected to Nitschkioideae Fitzp., the rank and author citation remaining unchanged.]

Recommendation 19A

If a legitimate name is not available for a taxon of a rank below family and above genus which includes the type genus of the name of another higher or lower taxon (e.g., subfamily, tribe, or subtribe), but not that of the family to which it is assigned, the new name of that taxon should be based on the same generic name as the name of the higher or lower taxon.

Section 3. NAMES OF GENERA AND SUBDIVISIONS OF GENERA
Article 20

The name of a genus is a substantive in the singular number, or a word treated as such. It may be taken from any source whatever, and may even be composed in an absolutely arbitrary manner.

[EXAMPLES: *Alectoria* Ach. (see p. 40), *Helicosporium* Nees ex Fr. (*helic*—coiled; *spora*—spore), *Hemimycena* Sing. (ἥμι—half; the genus *Mycena*), *Oudemansia* Speg. (after C. A. J. A. Oudemans), *Padixonia* Subram. (after P. A. Dixon), *Rinomia* Nieuw. (anagram of *Morinia*), *Seimatosporiopsis* Sutton *et al.* (*Seimatosporium*; -*opsis*—like).]

The name of a genus may not coincide with a technical term currently used in morphology unless it was published before 1 Jan. 1912 and was accompanied, when originally published, by a specific name published in accordance with the binary system of Linnaeus.

[EXAMPLES: If any generic names such as '*Basidium*', '*Conidia*', '*Pileus*', or '*Soralia*' were proposed after this date they would be contrary to this Article and rejected. *Isidium* (Ach.)Ach. (*Meth. Lich.*: xxxiii, 136, 1803) and *Tuber* Mich. ex Fr. (*Syst. mycol.* **2**: 289, 1823), however, appeared before this date, were accompanied by binomials when validly published, and so are admissible.]

The name of a genus may not consist of two words, unless these words are joined by a hyphen.

The following are not to be regarded as generic names:

(1) Words not intended as names.

(2) Unitary designations of species.

[EXAMPLES: Ehrhart proposed unitary names for species then known by binary names, e.g. *Scalopodora* Ehrh. for *Lichen velleus* L. (*Phytophylacium*, no. 80, 1780). These names, which resemble generic names, are not to be confused with them and must be rejected unless they have been taken up as generic names and validly published and used in the binary system by a later author. Necker proposed unitary names for his 'species naturales' (e.g. *Inodisum* Neck., *Elementa Botanica*: 348, 1790), which also resemble generic names and were catalogued as such by Zahlbruckner (*Cat. Lich. Univ.*, 1–10, 1921–40), but must be rejected unless taken up and validly published as generic names in the binary system by a later author. When such names are validated by a later author only the citation of the validating author is required as the unitary names are inadmissible.]

Recommendation 20A

Botanists who are forming generic names should comply with the following suggestions:

(a) To use Latin terminations insofar as possible.

(b) To avoid names not readily adaptable to the Latin language.

(c) Not to make names which are very long or difficult to pronounce in Latin.

(d) Not to make names by combining words from different languages.

(e) To indicate, if possible, by the formation or ending of the name the affinities or analogies of the genus.

(f) To avoid adjectives used as nouns.

(g) Not to use a name similar to or derived from the epithet of one of the species of the taxon.

(h) Not to dedicate genera to persons quite unconnected with botany or at least with natural science. [See also *Taxon* **10**: 214–221, 1961.]

(i) To give a feminine form to all personal generic names, whether they commemorate a man or a woman (see Rec. 73B).

(j) Not to form generic names by combining parts of two existing generic names, e.g. *Hordelymus* from *Hordeum* and *Elymus*, because such names are likely to be confused with names of intergeneric hybrids.

Article 21

The name of a subdivision of a genus is a combination of a generic name and a subdivisional epithet connected by a term (subgenus, section, series, etc.) denoting its rank.

The epithet is either of the same form as a generic name, or a plural adjective agreeing in gender with the generic name and written with a capital initial letter.

The epithet of a subgenus or section is not to be formed from the name of the genus to which it belongs by adding the ending *-oides* or *-opsis*, or the prefix *Eu-*.

[EXAMPLES: *Alectoria* subgen. *Bryopogon* Th. Fr., *Cladonia* subgen. *Cenomyce* sect. *Chasmariae* (Ach.) Flörke, and *Lecidea* subgen. *Lecidea* sect. *Acrocyaneae* Hertel are in accord with this Article, whereas *Alectoria* subgen. *Eualectoria* Th. Fr. is not (and must be changed to *A.* subgen. *Alectoria* without any author citation; see Art. 22).]

Note 1. The use within the same genus of the same epithet for subdivisions of the genus, even if they are of different rank, based on different types is illegitimate under Art. 64.

Recommendation 21A

When it is desired to indicate the name of a subdivision of the genus to which a particular species belongs in connection with the generic name and specific epithet, its epithet is placed in parentheses between the two; when necessary, its rank is also indicated.

[EXAMPLES: When originally described the two following species names were cited as *Cladonia* (ser. *Clausae, Thallostelides*) *subisabellina* des Abb. (*Rev. bryol. lichén.* **34**: 821, 1967), and *Parmelia* (*Amphigymnia*) *hendrickxii* Dodge (*Beih. Nova Hedwigia* **38**: 48, 1971).]

Recommendation 21B

The epithet of a subgenus or section is preferably a substantive, that of a subsection or lower subdivision of a genus preferably a plural adjective.

Botanists, when proposing new epithets for subdivisions of genera, should avoid those in the form of a substantive when other co-ordinate subdivisions of the same genus have them in the form of a plural adjective, and vice-versa.

They should also avoid, when proposing an epithet for a subdivision of a genus, one already used for a subdivision of a closely related genus, or one which is identical with the name of such a genus.

If it is desired to indicate the resemblance of a subgenus or section (other than the type subgenus or section) of one genus to another genus, the ending *-oides* or *-opsis* may be added to the name of that other genus to form the epithet of the subgenus or section concerned.

Article 22

The subgenus or section (but not subsection or lower subdivision) including the type species of the correct name of the genus to which it is assigned bears that generic name unaltered as its epithet, but without citation of an author's name (see Art. 46). The type of the correct name of each such subgenus or section is the same as that of the generic name. This provision does not apply to sections which include the type species of the names of other subgenera of the genus. The names of such sections are subject to the provisions of priority; they may repeat the name of the subgenus if no other epithet is available (see Rec. 22A).

The first valid publication of a name of a subgenus or section which does not include the type of the correct name of the genus automatically establishes the name of another subgenus or section respectively which does include that type and which bears as its epithet the generic name unaltered. Such autonyms (automatically established names) are not to be taken into consideration for purposes of priority. However, when no other epithet is available, the epithets of autonyms may be adopted as new in another position or rank.

[EXAMPLES: The section of *Ascobolus* Pers. ex Hook. which includes the nomenclatural type of that genus (*A. furfuraceus* Pers. ex Hook.) is called sect. *Ascobolus*. The correct name for the section of *Alectoria* which includes *A. pubescens* (L.) R. Howe (not the nomenclatural type of *Alectoria*) is the earliest legitimate name in the rank of section, i.e. sect. *Teretiuscula* (Hillm.) Lamb, and not sect. *Pseudephebe* (Choisy) Choisy nor sect. *Subparmelia* Degel. In describing *Fulgensia* subgen. *Candelariopsis* (Ceng. Samb.) Poelt (*Mitt. Bot. StSamml., Münch.* **5**: 594, 1965) which is typified by a species other than the type of the generic name *Fulgensia*, *Fulgensia* subgen. *Fulgensia* was automatically established (typified by the type species of the generic name).]

The final epithet in the name of a subdivision of a genus may not repeat unchanged the correct name of the genus, except when the two names have the same type.

When the epithet of a subdivision of a genus is identical with or derived from the epithet of one of its constituent species, this species is the type of the name of the subdivision of the genus unless the original author of that name designated another type.

[EXAMPLE: When *Lecidea* sect. *Elaeochromae* subsect. *Carpathicae* Poelt was described (*Ber. Bayer. Bot. Ges.* **34**: 84, 1961) the author did not definitely state the type species but this name must be typified by one of the constituent species from which the subsectional name was derived (i.e. *L. carpathica* (Körb.) Szat).]

Note. When the epithet of a subdivision of a genus is identical with or derived from the epithet of a specific name that is a later homonym, it is the species designated by that later homonym, whose correct name necessarily has a different epithet, that is the nomenclatural type.

Recommendation 22A

A section including the type of the correct name of a subgenus, but not including the type of the correct name of the genus, should, where there is no obstacle under the rules, be given the same epithet and type as the subgenus.

A subgenus not including the type of the correct name of the genus should, where there is no obstacle under the rules, be given the same epithet and type as one of its subordinate sections.

[EXAMPLE: Instead of using a new name at the subgeneric level Hale (in Hale and Kurokawa, *Contr. U.S. natn. Herb.* **36**: 121, 1964) raised *Parmelia* sect. *Xanthoparmelia* Vain. to the rank of subgenus as *P.* subgen. *Xanthoparmelia* (Vain.) Hale. The type species of both names is the same (i.e. *Parmelia conspersa* (Ach.) Ach.).]

Section 4. NAMES OF SPECIES

Article 23

The name of a species is a binary combination consisting of the name of the genus followed by a single specific epithet. If an epithet consists of two or more words, these are to be united or hyphened. An epithet not so joined when originally published is not to be rejected but, when used, is to be united or hyphened.

The epithet of a species may be taken from any source whatever, and may even be composed arbitrarily.

[EXAMPLES: *Penicillium flavoglaucum* Biourge, *Verticillium albo-atrum* Reinke & Berthold. '*Ochrolechia elisabethae kolae*' Vers. has to be changed to *O. elisabethae-kolae* Vers.]

The specific epithet may not exactly repeat the generic name with or without the addition of a transcribed symbol (tautonym).

[EXAMPLES: *Cyathus crucibulum* Mich. ex Pers. cannot be used as the correct name for *Crucibulum vulgare* Tul. when it is treated as a member of the genus *Crucibulum* Tul., although it predates the epithet *vulgare* at species rank by 43 years, as the resultant binomial would be a tautonym, i.e. '*Crucibulum crucibulum*'. *Acetabularia acetabulosa* Massee and *Dubitatio dubitationum* Speg. are, however, acceptable.]

The specific epithet, when adjectival in form and not used as a substantive, agrees grammatically with the generic name.

[EXAMPLES: *Agaricus melleus* Vahl ex Fr. or *Armillaria mellea* (Vahl ex Fr.) Kumm. (adjectival epithet), *Ganoderma lucidum* (Curt. ex Fr.) Karst. (adjectival), *G. colossus* (Fr.) C. Baker (substantive so not '*colossum*') and *Chaetomium funicola* Cooke ('-*icola*' is an invariable Latin substantive so not '*funicolum*'). See also Stearn (1973).]

The following are not to be regarded as specific epithets:

(1) Words not intended as names.

(2) Ordinal adjectives used for enumeration.

[EXAMPLES: *Boletus vicesimus sextus*, *Agaricus octogesimus nonus*.]

(3) Epithets published in works in which the Linnaean system of binary nomenclature for species is not consistently employed.

[NOTE: Point (3) refers mainly to polynomial names (descriptive phrases) used by most pre-Linnean authors but also by a few later (mainly eighteenth century) authors; any such names which consist of only two words are not considered to be in the Linnean binomial system and rejected. The fungus names of Secretan (*Mycol. suisse*, 1833) are also affected by this Article (see *Taxon* **11**: 170–173, 1962).]

Recommendation 23A

Names of men and women and also of countries and localities used as specific epithets may be substantives in the genitive (*clusii, saharae*) or adjectives (*clusianus, dahuricus*) (see also Art. 73).

It will be well, in the future, to avoid the use of the genitive and the adjectival form of the same word to designate two different species of the same genus; for example, *Lysimachia hemsleyana* Maxim. (1891) and *L. hemsleyi* Franch. (1895).

Recommendation 23B

In forming specific epithets, botanists should comply also with the following suggestions:

(a) To use Latin terminations insofar as possible.

(b) To avoid epithets which are very long and difficult to pronounce in Latin.

(c) Not to make epithets by combining words from different languages.

(d) To avoid those formed of two or more hyphened words.

(e) To avoid those which have the same meaning as the generic name (pleonasm).

(f) To avoid those which express a character common to all or nearly all the species of a genus.

(g) To avoid in the same genus those which are very much alike, especially those which differ only in their last letters or in the arrangement of two letters.

(h) To avoid those which have been used before in any closely allied genus.

(i) Not to adopt unpublished names found in correspondence, travellers' notes, herbarium labels, or similar sources, attributing them to their authors, unless these authors have approved publication.

(j) To avoid using the names of little-known or very restricted localities, unless the species is quite local.

[EXAMPLES: Epithets not recommended: *Heterochaete ogasawarasimensis* Ito & Imai, *Mukagomyces hiromichii* Imai, *Aspicilia thjanschanica* Oksn., *Stereogloeocystidium subsanguinolentum* Rick (see also p. 42). The herbarium name '*Sticta latiloba* Tayl.' should not have been taken up and published as *Pseudocyphellaria lobata* Tayl. ex Dodge (*Nova Hedwigia* **19**: 491, 1971) as Taylor died in 1848 and did not authorize its publication 123 years after his death!]

Section 5. NAMES OF TAXA BELOW THE RANK OF SPECIES (INFRASPECIFIC TAXA)

Article 24

The name of an infraspecific taxon is a combination of the name of a species and an infraspecific epithet connected by a term denoting its rank. Infraspecific epithets are formed as those of species and, when adjectival in form and not used as substantives, they agree grammatically with the generic name.

Infraspecific epithets such as *typicus, originalis, originarius, genuinus, verus,* and *veridicus,* purporting to indicate the taxon containing the nomenclatural type of the next higher taxon, are inadmissible and cannot be validly published except where they repeat the specific epithet because Art. 26 requires their use.

The use of a binary combination for an infraspecific taxon is not admissible.

[EXAMPLES: *Stereocaulon denudata* subsp. *santorinense* Stein (not '*Stereocaulon santorinense* Stein' as listed by Zahlbruckner, *Cat. Lich. Univ.* 4: 666, 1927). Names in the form '*Alectoria ochroleuca* **A. thulensis* Th. Fr.' are to be changed to *A. ochroleuca* subsp. *thulensis* Th. Fr. *Pertusaria coccodes* f. *albocincta* (Erichs.) Grumm. may be cited as *P. coccodes* var. *phymatodes* f. *albocincta* if the full classification of the form within the species is required. The infraspecific epithet '*typica*' in *Cetraria crispa* var. *typica* Savicz is inadmissible as it was intended to include the type of *C. crispa* (Ach.) Nyl. (i.e. should have been referred to as 'var. *crispa*').]

Infraspecific taxa within different species may bear the same epithets; those within one species may bear the same epithets as other species (but see Rec. 24B).

[EXAMPLES: *Fusarium aquaeductuum* var. *majus* Wollenw. and *Fusarium arcuatum* var. *majus* Wollenw. are both acceptable; *Cladonia furcata* f. *subulata* (Ach.) Vain. is acceptable in spite of the existence of *C. subulata* (L.) Wigg. which is based on a different type.]

Note. The use within the same species of the same epithet for infraspecific taxa, even if they are of different rank, based on different types is illegitimate under Art. 64.

Recommendation 24A

Recommendations made for specific epithets (see Recs. 23A, B) apply equally to infraspecific epithets.

Recommendation 24B

Botanists proposing new infraspecific epithets should avoid those previously used for species in the same genus.

Article 25

For nomenclatural purposes, a species or any taxon below the rank of species is regarded as the sum of its subordinate taxa, if any.

Article 26

The name of an infraspecific taxon which includes the type of the correct name of the species has as its final epithet the same epithet, unaltered, as that of the correct name of the species, but without citation of an author's name (see Art. 46). The type of the correct name of each such infraspecific taxon is the same as that of the correct name of the species. If the epithet of the species is changed, the names of those infraspecific taxa which include the type of the name of the species are changed accordingly.

[EXAMPLE: The infraspecific taxon *Lobaria ferax* var. *genuina* Vain. included the type of the species *L. ferax* Vain. so the varietal epithet must be changed to var. *ferax*. Linnaeus (*Sp. Pl.* **2**: 1153, 1753) recognized two varieties of *Lichen rangiferinus* L., var. *alpestris* L. and var. *sylvaticus* L., one of which must have included the type of the species name; in this case var. *sylvaticus* has to be changed to var. *rangiferinus* as it is considered to be the element including the type of *L. rangiferinus* (see *Taxon* **15**: 64–66, 1966). See also example under Art. 24.]

The first valid publication of a name of an infraspecific taxon which does not include the type of the correct name of the species automatically establishes the name of a second taxon of the same rank which does include that type and has the same epithet as the species. Such autonyms (automatically established names) are not to be taken into consideration for purposes of priority. Where no other epithet is available, the epithets of autonyms may be adopted as new in another position or rank.

[EXAMPLE: When Booth (*Mycol. Pap.* **83**: 7, 1961) made the new combination *Thielavia terricola* var. *minor* (Rayss & Borut) C. Booth he automatically established *T. terricola* var. *terricola*, while Rayss and Borut (*Mycopathologia* **10**: 160, 1959) in describing *T. terricola* f. *minor* Rayss & Borut, had previously automatically established a f. *terricola*. Neither var. *terricola* nor f. *terricola* are given author citations.]

Recommendation 26A

A variety including the type of the correct name of a subspecies, but not including the type of the correct name of the species, should, where there is no obstacle under the rules, be given the same epithet and type as the subspecies. A subspecies not including the type of the correct name of the species should, where there is no obstacle under the rules, be given the same epithet and type as one of its subordinate varieties.

A taxon of lower rank than variety which includes the type of the correct name of a subspecies or variety, but not the type of the correct name of the species, should, where there is no obstacle under the rules, be given the same epithet and type as the subspecies or variety. On the other hand, a subspecies or variety which does not include the type of the correct name of the species should not be given the same epithet as that of one of its subordinate taxa below the rank of variety.

[EXAMPLE: As no legitimate epithet was available in the rank of subspecies Laundon (*Lichenologist* **2**: 63, 1962) made the new combination *Bacidia citrinella* subsp. *alpina* (Schaer.) Laund. using the same epithet used in 1902 by Boistel in making the combination *B. citrinella* var. *alpina* (Schaer.) Boist. (both based on *Lecidea flavovirescens* var. *alpina* Schaer.) and did not introduce a new name in subspecies rank.]

Article 27

The final epithet in the name of an infraspecific taxon may not repeat unchanged the epithet of the correct name of the species to which the taxon is assigned except when the two names have the same type.

Chapter IV. EFFECTIVE AND VALID PUBLICATION
Section 1. CONDITIONS AND DATES OF EFFECTIVE PUBLICATION

Article 29

Publication is effected, under this Code, only by distribution of printed matter (through sale, exchange, or gift) to the general public or at least to botanical institutions with libraries accessible to botanists generally. It is not effected by communication of new names at a public meeting, by the placing of names in collections or gardens open to the public, or by the issue of microfilm made from manuscripts, type-scripts or other unpublished material. Offer for sale of printed matter that does not exist does not constitute effective publication.

Publication by indelible autograph before 1 Jan. 1953 is effective.

Note. For the purpose of this Article, handwritten material, even though reproduced by some mechanical or graphic process (such as lithography, offset, or metallic etching), is still considered as autographic.

Publication on or after 1 Jan. 1953 of a new name in tradesmen's catalogues or non-scientific newspapers, and on or after 1 Jan. 1973 in seed-exchange lists, does not constitute effective publication.

Recommendation 29A

Authors are urged to avoid publishing new names or descriptions in ephemeral publications, in popular periodicals, in any publication unlikely to reach the general botanical public, in those produced by such methods that their permanence is unlikely, or in abstracting journals.

Article 30

The date of effective publication is the date on which the printed matter became available as defined in Art. 29. In the absence of proof establishing some other date, the one appearing in the printed matter must be accepted as correct.

[EXAMPLE: The generic name *Chrysocyclus* Syd. was published in a journal dated 31 December 1925 while *Holwayella* Jacks. appeared in one dated one day later (1 January 1926). When these genera are considered synonymous *Chrysocyclus* has priority (Art. 11) as there is no proof that these dates are wrong. For further information on ascertaining dates of publication see pp. 110–111.]

When separates from periodicals or other works placed on sale are issued in advance, the date on the separate is accepted as the date of effective publication unless there is evidence that it is erroneous.

[EXAMPLES: Separates from Karsten's paper in *Medd. Soc. Fauna Fl. fenn.* **5**: 1–48 (1880) are known to have been issued in 1879 (see *Rev. mycol.* **2**: 136, 1880) and so the new names proposed in that date from 1879 not 1880. Th. Fries' work in *Nova Acta R. Soc. Scient. upsal., ser.* 3, **3**: 103–398 (1861) was issued in advance in the form of separates with changed pagination and dated '1860' and so is considered to have been effectively published in 1860 and not 1861.]

Recommendation 30A

The date on which the publisher or his agent delivers printed matter to one of the usual carriers for distribution to the public should be accepted as its date of publication.

Article 31

The distribution on or after 1 Jan. 1953 of printed matter accompanying exsiccata does not constitute effective publication.

Note. If the printed matter is also distributed independently of the exsiccata, this constitutes effective publication.

[EXAMPLE: Works such as Lundell & Nannfeldt's *Fungi exsiccati suecici* (Helsingfors) **1**→ (1934→) and Vězda's *Lichenes selecti exsiccati* (Brno) **1**→ (1960→) in which the printed labels are issued independently of the dried specimens are effectively published. Names in exsiccata distributed prior to 1 January 1953 meeting the requirements of Art. 29 are effectively published whether printed matter was issued independently or not.]

Section 2. CONDITIONS AND DATES OF VALID PUBLICATION OF NAMES

Article 32

In order to be validly published, a name of a taxon must (1) be effectively published (see Art. 29); (2) have a form which complies with the provisions of Arts. 16–27* (but see Art. 18, notes 1, 2); (3) be accompanied by a description or diagnosis** of the taxon or by a reference (direct or indirect) to a previously and effectively published description or diagnosis of it; and (4) comply with the special provisions of Arts. 33–45.

Note 1. Names published with an incorrect Latin termination but otherwise in accordance with this Code are regarded as validly published; they are to be changed to accord with Arts. 17–19, 21, 23, and 24, without change of the author's name.

Note 2. An indirect reference is a clear indication, by the citation of the author's name or in some other way, that a previously and effectively published description or diagnosis applies to the taxon to which the new name is given.

[EXAMPLES: The lichen genus *Reticularia* Baumg. (*Fl. lips.*: 569, 1790) was published without a description or diagnosis or a reference to a former one so is not validly published and cannot affect the nomenclatural status of the myxomycete genus *Reticularia* Bull. (*Hist. Champ. Fr.*: 83, 1791) based on a different type. The species name *Cetraria ericetorum* Opiz (*Seznam rostlin'květeny české*: 175, 1852) was validly published, however, although no description or diagnosis was provided, as *C. islandica* var. *crispa* Ach. (a validly published name) was indicated to be a replaced synonym. The generic name *Zygospermella* Cain (*Mycologia* 27: 227, 1935), published with no description or diagnosis, is validly published as it refers to the previously validly published description of *Zygospermum* Cain (*Univ. Toronto Stud., biol. ser.* 38: 73, 1934).]

Note 3. In certain circumstances an illustration with analysis is accepted as equivalent to a description (see Arts. 42 and 44).

[EXAMPLE: The generic name *Ravenelia* Berk. (*Gdners Chron.* 10: 132, 1853) is validly published without a formal description as the figures and comments are equivalent to a diagnosis and description.]

* [As '16–17' in Stafleu *et al.* (1972) but see *Taxon* 22: 503 (1973).]
** A diagnosis of a taxon is a statement of that which in the opinion of its author distinguishes the taxon from others.

Note 4. For names of plant taxa originally published as names of animals, see Art. 45.

Recommendation 32A

Publication of a name should not be validated solely by a reference to a description or diagnosis published before 1753.

[NOTE: Names which have been validated in the past by pre-starting point references are to be accepted but authors who are introducing new names should not follow this practice (though their names would also be accepted).]

Recommendation 32B

The description or diagnosis of any new taxon should mention the points in which the taxon differs from its allies.

Recommendation 32C

Authors should avoid adoption of a name or an epithet which has been previously but not validly published for a different taxon.

Recommendation 32D

In describing new taxa, authors should, when possible, supply figures with details of structure as an aid to identification.

In the explanation of the figures, it is valuable to indicate the specimen(s) on which they are based.

Authors should indicate clearly and precisely the scale of the figures which they publish.

Recommendation 32E

The description or diagnosis of parasitic plants should always be followed by an indication of the hosts, especially those of parasitic fungi. The hosts should be designated by their scientific names and not solely by names in modern languages, the applications of which are often doubtful.

Article 33

A combination is not validly published unless the author definitely indicates that the epithet or epithets concerned are to be used in that particular combination.

[EXAMPLES: Definite indication by the author usually means applying and accepting the combination as the correct one for the taxon concerned. In a discussion of alternative taxonomic treatments of *Verticillium dahliae* Kleb., for example, Isaac (*Trans. Br. mycol. Soc.* **32**: 154, 1949) mentioned the new combination '*V. albo-atrum* var. *dahliae*' but is not considered to have validly published this name as he accepted *V. dahliae* as the correct name for the taxon. In raising *Siphula* sect. *Siphulina* Hue (nom. inval., Art. 32) to generic rank as *Siphulina* Dodge, Dodge (*Trans. Am. microsc. Soc.* **84**: 510, 1965) cited '*Siphula orphina* Hue' as the nomenclatural type of his name, but as he did not transfer the epithet '*orphina*' into a new combination clearly he is regarded as not having validly published a combination of this epithet under *Siphulina*.]

A new combination or a new name for a previously recognized taxon published on or after 1 Jan. 1953 is not validly published unless its basionym (name-bringing or epithet-bringing synonym) or the replaced synonym (when a new name or epithet is proposed) is clearly indicated and a full and direct reference given to its author and original publication with page or plate reference and date.

[EXAMPLES: The combination '*Hypoxylon necatrix* (Hartig) P. Martin' (*Jl S. Afr. Bot.* **34**: 187, 1968) was not validly published as the full reference to the place of publication of the basionym was omitted. *Phialophora repens* (Davidson) Conant (*Mycologia* 29: 598, 1937), however, was validly published as although the basionym was not cited the name was published before 1953. The new name *Alectoria furcellata* R. Sant. (*Svensk bot. Tidskr.* 62: 489, 1968) and the new combination *Thelotrema subdenticulatum* (Zahlbr.) G. Salisb. (*Lichenologist* 5: 267, 1972) are both validly published as they include full and direct references to the replaced synonym and basionym, respectively.]

Note 1. Mere reference to the *Index Kewensis*, the *Index of Fungi*, or any work other than that in which the name was validly published does not constitute a full and direct reference to the original publication of a name.

[EXAMPLES: Ciferri (*Mycopath. mycol. appl.* 7: 86–89, 1954), in making some 142 new combinations into *Meliola*, omitted all references to the place of publication of the basionyms stating they could be found through Petrak's lists and the *Index of Fungi*; all these combinations are treated as not validly published. Grummann (*Cat. Lich. Germ.*: 18, 1963) introduced a new combination in the form '*Lecanora campestris* f. *pseudistera* (Nyl.) Grumm.c.n. — *L.p.* Nyl., Z5: 521', in which 'Z' refers to Zahlbruckner's *Catalogus lichenum universalis* (see p. 108) in which the full citation of Nylander's name is given. Grummann's combination is treated as not validly published under this Note.]

Note 2. Bibliographic errors of citation do not invalidate the publication of a new combination.

[EXAMPLE: Poelt and Follmann, in introducing the new combination *Dirina stenhammari* (*Herzogia* 1: 64, 1968), cited *Lecanactis stenhammari* Fr. ex Arnold (1871) as the basionym; Fries' epithet was later found to have been validly published in 1848 as *Lecidea stenhammari* Fr. and is the correct basionym. The combination *Dirina stenhammari* is still treated as validly published in 1968 but its author citation has to be changed to '(Fr.) Poelt & Follm.' instead of '(Fr. ex Arnold) Poelt & Follm.' (see *Philippia* 1: 127–128, 1972).]

A name given to a taxon whose rank is at the same time denoted by a misplaced term (one contrary to Art. 5) is treated as not validly published, examples of such misplacement being a form divided into varieties, a species containing genera, or a genus containing families or tribes.

An exception is made for names of the infrageneric taxa termed tribes (*tribus*) in Fries' *Systema Mycologicum*, which are treated as validly published.

[EXAMPLES: Lloyd (*Mycol. Notes* 3: 170, 1912) used the new name *Lentus* Lloyd giving it the rank of 'section' although he employed the names as if they were generic in binomials (e.g. '*L. tricholoma*'). Because his names were given a misplaced term they are treated as not validly published (see *Persoonia* 1: 233–35, 1960). The name *Agaricus* trib. *Armillaria* Fr. ex Fr. (*Syst. mycol.* 1: 26, 1821) is treated as validly published, however, and so can be raised to the rank of genus (cf. p. 137).]

Article 34

A name is not validly published (1) when it is not accepted by the author in the original publication; (2) when it is merely proposed in anticipation of the future acceptance of the group concerned, or of a particular circumscription, position, or rank of the group (so-called provisional name); (3) when it is merely mentioned incidentally; (4)

when it is merely cited as a synonym; (5) by the mere mention of the subordinate taxa included in the taxon concerned.

Note 1. Provision no. 1 does not apply to names or epithets published with a question mark or other indication of taxonomic doubt, yet published and accepted by the author.

Note 2. By 'incidental mention' of a new name or combination is meant mention by an author who does not intend to introduce the new name or combination concerned.

[EXAMPLES: The examples are numbered according to the criteria in the first paragraph of this Article. (1) *Alectoria karelica* Räs. (*Ann. Bot. Soc. zool.-bot. fenn. Vanamo* **12**(1): 34, 1939) was not accepted by Räsänen when first published as he concluded it was a synonym of another taxon. (2) & (3) Saccardo and Sydow (*Syll. Fung.* **14**: 787, 1899) suggested the new generic name *Schnablia* Sacc. & Syd. for *Belonidium schnablianum* Rehm but did not adopt *Schnablia* as the name for this species and so did not validly publish it; *Chaetomiotricha* Peyr. (*Annls mycol.* **12**: 462, 1914) and Asterothyriaceae W. Wats. (*New Phytol.* **28**: 33, 1929) are invalid for similar reasons. (2) *Lecidea "pseudassimilis"* Hertel (*Herzogia* **1**: 426, 1970) was introduced as '(nom. provis.)' so although it was accepted in that work it is still not validly published. (4) *Nephromium servitianum* (Gyeln.) Gyeln. (*Annls Mus. Nat. Hungar.* **28**: 279, 1934) is not validly published as it was listed as a synonym of *Nephroma servitianum* Gyeln. (5) The generic name *Stenocybe* Nyl. (*Bot. Notiser* **3**: 84, 1854) was introduced as '−−*Stenocybe* Nyl. speciebus *St. byssacea* (Fr.) et *St. majore* Nyl. (in Zw. exs. 71)−−' without any generic description and so is not validly published; this name dates from Körber (*Syst. lich. Germ.*: 306, 1855) who published the name in accordance with Art. 32 and is cited as *Stenocybe* Nyl. ex Körb. From Note 1, however, names such as '*Apiosporium* ? *erysiphoides* Sacc. & Ellis' (*Michelia* **2**: 566, 1882) which were accepted by their publishing authors are treated as valid.

The 'pseudogeneric names' of the fictitious Prof. N. J. McGinty coined by Lloyd jocularly are also treated as invalid under this Article (see *Reintwardia* **1**: 205, 1951).]

When, on or after 1 Jan. 1953, two or more different names (so-called alternative names) are proposed simultaneously for the same taxon by the same author, none of them is validly published (but see Art. 59).

[EXAMPLES: Murrill (*Q. J. Fla Acad. Sci.* **8**: 175–198, 1945) described a number of new species and at the end of the paper made combinations transferring fourteen of these into other genera 'for those using Saccardo'; as both appear in the same work these are 'alternative names', but as they appeared before 1 January 1953 both the new species and combinations are validly published, although such names published on or after this date are not validly published. Paul (*Trans. Br. mycol. Soc.* **56**: 261, 263, 1971) simultaneously published the names *Pyrenophora tetrarrhenae* A. Paul and *Drechslera tetrarrhenae* A. Paul for the same fungus; as these refer to the perfect and imperfect states, respectively, they are not alternative names in the sense of this Article and so both are validly published.]

Recommendation 34A

Authors should avoid publishing or mentioning in their publications unpublished names which they do not accept, especially if the persons responsible for these names have not formally authorized their publication (see Rec. 23B, i).

[EXAMPLE: Gilbert (in Bresadola, *Icon. mycol.* **27**: 73, 1940) cited '*Gilbertia* Donk, in litteris 1934' as a synonym of his new genus *Aspidella* E. Gilb. As this name had not appeared in print elsewhere and Donk had not authorized its publication it should not have been mentioned by Gilbert. For a suggested treatment of unpublished names see p. 67 and Fig. 6. See also under Rec. 23B.]

Article 35

A new name published on or after 1 Jan. 1953 without a clear indication of the rank of the taxon concerned is not validly published.

For such names published before 1 Jan. 1953 the choice made by the first author who assigned a definite rank is to be followed.

Article 36

In order to be validly published, a name of a new taxon of plants, the bacteria, algae, and all fossils excepted, published on or after 1 Jan. 1935 must be accompanied by a Latin description or diagnosis or by a reference to a previously and effectively published Latin description or diagnosis of the taxon.

[EXAMPLES: *Ancylospora* Saw. (*Rep. Govt. Res. Inst. Formosa* **87**: 77, 1944), *Lactarius volemus* f. *gracilis* Heim (*Rev. mycol.* **27**: 143, 1962) and *Venturia acerina* Plakidas (*Mycologia* **34**: 34, 1942) are not validly published as they were published without Latin diagnoses after 1 Jan. 1935 having descriptions in Japanese, French and English, respectively. *Herpotrichia nicaraguensis* Ellis & Everh. (*Bull. Lab. nat. hist. Univ. Iowa* **2**: 400, 1893), however, is validly published as although published only with an English diagnosis it appeared before 1 Jan. 1935. *Chaetomium osmaniae* Rama Rao & Reddy (*Mycopath. mycol. appl.* **31**: 74, 1967) is validly published as it is a new name to replace the illegitimate *C. elongatum* Rama Rao & Reddy described in 1964 with a Latin diagnosis.]

Recommendation 36A

Authors publishing names of new taxa of recent plants should give or cite a full description in Latin in addition to the diagnosis.

Article 37

Publication on or after 1 Jan. 1958 of the name of a new taxon of the rank of family or below is valid only when the nomenclatural type is indicated (see Arts. 7–10).

[EXAMPLES: *Leptogium hibernicum* Mitchell (*Bull. Soc. Sci. Bretagne* **37**: 119, 1964) and *Phaeographis polymorpha* Nakanishi (*J. Sci. Hiroshima Univ., ser.B, div. 2*, **11**: 79, 1966) are not validly published here as although several collections were cited with the original descriptions a holotype was not designated in these cases (see also under Art. 45). In describing *Alectoria canadensis* Mot., Motyka (*Bryologist* **67**: 35, 1964) stated that the type of the name was in H but listed more than one collection from that herbarium, not specifying which was the type, and so this name is also invalid. In describing the new genus *Fuscidea*, Wirth and Vězda (*Diss. Bot.* **17**: 286, 1972) included 15 species within it but did not designate a holotype species and so did not validly publish the new generic name; this name was validated in a paper published later in the same year (*Beitr. naturk. Forsch. SudwDtl.* **31**: 91, Nov. 1972) when a holotype was designated.]

Recommendation 37A

The indication of the nomenclatural type should immediately follow the Latin description or diagnosis and should be given by the insertion of the Latin word 'typus' (or 'holotypus', etc.) immediately before or after the particulars of the type so designated.

Recommendation 37B

When the nomenclatural type of a new taxon is a specimen, the place where it is permanently conserved should be indicated.

Article 38

In order to be validly published, a name of a new taxon of fossil plants of specific or lower rank published on or after 1 Jan. 1912 must be accompanied by an illustration or figure showing the essential characters, in addition to the description or diagnosis, or by a reference to a previously and effectively published illustration or figure.

Article 41

In order to be validly published, a name of a genus must be accompanied (1) by a description or diagnosis of the genus, or (2) by a reference (direct or indirect) to a previously and effectively published description or diagnosis of the genus in that rank or as a subdivision of a genus.

Note. In certain circumstances, an illustration with analysis is accepted as equivalent to a generic description (see Art. 42).

[EXAMPLES: The following generic names were validly published for the reasons indicated: *Absconditella* Vězda (*Preslia* **37**: 238, 1965), accompanied by a generic description; *Orceolina* Hertel (*Vortr. GesGeb. Bot.* **4**: 182, 1970) by reference to the previously validly published but illegitimate name *Urceolina* Tuck.; and *Tetramelaena* (Trevis.) Dodge (*Beih. Nova Hedwigia* **38**: 200, 1971) based on the validly published name *Dimelaena* sect. *Tetramelaena* Trevis.]

Article 42

The publication of the name of a monotypic new genus based on a new species is validated either by (1) the provision of a combined generic and specific description (*descriptio generico-specifica*) or diagnosis, or (2), for generic names published before 1 Jan. 1908, by the provision of an illustration with analysis showing essential characters (see Art. 32, Note 3).

[EXAMPLE: The monotypic genus *Ingoldiella* D. Shaw (*Trans. Br. mycol. Soc.* **59**: 258, 1972) was introduced as '*Ingoldiella hamata* gen. et sp. nov.' with a combined Latin description and is consequently validly published.]

A description or diagnosis of a new species assigned to a monotypic new genus is treated also as a generic description or diagnosis if the genus is not separately defined. However, the name of a monotypic genus of fossil plants published on or after 1 Jan. 1953 must be accompanied by a description or diagnosis of the genus.

A description or diagnosis of a monotypic new genus based on a new species is treated also as a specific description or diagnosis if the generic name and specific epithet are published together and the species is not separately defined.

Note. Single figures of microscopic plants showing the details necessary for identification are considered as illustrations with analysis showing essential characters.

Article 43

A name of a taxon below the rank of genus is not validly published unless the name of the genus or species to which it is assigned is validly published at the same time or was validly published previously.

[EXAMPLES: The species name *Gloeocercospora inconspicua* Demaree & Wilcox (*Phytopathology* **37**: 498, 1947) was not validly published, although provided with a Latin description and holotype, as the generic name *Gloeocercospora* Bain & Egerton (*Phytopathology* **33**: 225, 1943) was not validly published, having been described without a Latin diagnosis. This specific name correctly dates from 1971 when it and the generic name were validated by Deighton (*Trans. Br. mycol. Soc.* **57**: 358, 1971) and is cited as *G. inconspicua* Demaree & Wilcox ex Deight.

In 1880, Müller Argoviensis (*Flora, Jena* **63**: 286) published the new genus *Phlyctidia* with the species *P. hampeana* n. sp., *P. boliviensis* (= *Phlyctis boliviensis* Nyl.), *P. sorediiformis* (= *Phlyctis sorediiformis* Kremp.), *P. brasiliensis* (= *Phlyctis brasiliensis* Nyl.), and *P. andensis* (= *Phlyctis andensis* Nyl.). These specific names are, however, not validly published in this place, because the generic name *Phlyctidia* was not validly published; Müller gave no generic description or diagnosis but only a description and a diagnosis of the new species *P. hampeana*. This description and diagnosis cannot validate the generic name as a *descriptio generico-specifica* under Art. 42, since the new genus was not monotypic. The first valid publication of the name *Phlyctidia* was made by Müller in 1895 (*Hedwigia* **34**: 141), where a short generic diagnosis was given. The only species mentioned here were *P. ludoviciensis* n. sp. and *P. boliviensis* (Nyl.). The latter combination was validly published in 1895 by the reference to the basionym.]

Note. This Article applies also to specific and other epithets published under words not to be regarded as generic names (see Art. 20).

Article 44

The name of a species or of an infraspecific taxon published before 1 Jan. 1908 is validly published if it is accompanied only by an illustration with analysis showing essential characters (see Art. 32, Note 3).

Note. Single figures of microscopic plants showing the details necessary for identification are considered as illustrations with analysis showing essential characters.

[NOTE: See example cited under Art. 32 Note 3.]

Article 45

The date of a name or of an epithet is that of its valid publication. When the various conditions for valid publication are not simultaneously fulfilled, the date is that on which the last is fulfilled. A name published on or after 1 Jan. 1973 for which the various conditions for valid publication are not simultaneously fulfilled is not validly published unless a full and direct reference is given to the place or places where these requirements were previously fulfilled.

[EXAMPLES: The family name Dictyonemataceae Tom. (*Archo Bot. Sist. Fitogeogr. Genet.* **25**: 261, 1949) was not validly published under Art. 36 and dates from Tomaselli (*Ibid.* **26**: 104, 1950) when a Latin description was provided; the generic name *Gloeocercospora* dates from 1971 not 1943 (see example under Art. 43); the combination *Aspicilia obscurascens* (Magnusson) Clauz. & Rondon (*Ann. Soc. Hort. hist. nat. Hérault* **101**: 61, 1961) was not validly published under Art. 33 and dates from 1970 when Ozenda and Clauzade (*Les Lichens*: 775, 1970) gave the full bibliographic citation of the basionym; and the epithet '*subcana* Nyl.' in *Alectoria* was not validly published by Crombie (*J. Bot., Lond.* **14**: 360, 1876) under Art. 32 and correctly dates from 1892 when Stizenberger (*Ann. Naturhist. Mus. Wien* **7**: 129, 1892) provided a description of it (as *A. prolixa* var. *subcana*). Nakanishi (*J. Sci. Hiroshima Univ., ser. B, div. 2*, **11**: 51–126, 1966) failed to designate holotypes for the new species proposed with Latin descriptions so these were not validly published under Art. 37; these names correctly date from 1967 (*J. Sci. Hiroshima Univ., ser. B, div. 2*, **11**: 265, 1967) when holotypes were designated.]

Note. A correction of the original spelling of a name or epithet (see Art. 73) does not affect its date of valid publication.

For purposes of priority only legitimate names and epithets are taken into consideration (see Arts. 11, 63–67). However, validly published earlier homonyms, whether legitimate or not, shall cause rejection of their later homonyms (unless the latter are conserved).

If a taxon is transferred from the animal to the plant kingdom, its name or names available* under the International Code of Zoological Nomenclature and validly published in the form provided in the botanical Code shall be automatically accepted as having been validly published under this Code at the time of its valid publication as the name of an animal (see, however, Art. 65).

Recommendation 45A

Authors publishing a name of a new taxon in works written in a modern language (floras, catalogues, etc.) should simultaneously comply with the requirements of valid publication.

Recommendation 45B

Authors should indicate precisely the dates of publication of their works. In a work appearing in parts the last-published sheet of the volume should indicate the precise dates on which the different fascicles or parts of the volume were published as well as the number of pages and plates in each.

Recommendation 45C

On separately printed and issued copies of works published in a periodical, the date (year, month, and day), the name of the periodical, the number of its volume or parts, and the original pagination should be indicated.

Section 3. CITATION OF AUTHORS' NAMES
AND OF LITERATURE
FOR PURPOSES OF PRECISION

Article 46

For the indication of the name of a taxon to be accurate and complete, and in order that the date may be readily verified, it is necessary to cite the name of the author(s) who first validly published the name concerned unless the provisions of Arts. 19, 22, or 26 apply.

[EXAMPLES: Pleosporaceae Wint., *Pleospora* Rabenh., *Taphrina deformans* (Berk.) Tul., *Lecanora epibryon* var. *bryospora* Doppelbaur & Poelt.]

Recommendation 46A

Authors' names put after names of plants may be abbreviated, unless they are very short. For this purpose, particles are suppressed unless they are an inseparable part of the name, and the first letters are given without any omission (Lam. for J.B.P.A. Monet Chevalier de Lamarck, but De Wild. for É. De Wildeman).

If a name of one syllable is long enough to make it worth while to abridge it, the first consonants only are given (Fr. for Elias Magnus Fries); if the name has two or more syllables, the first syllable and the first letter of the following one are taken, or the two first when both are consonants (Juss. for Jussieu, Rich. for Richard).

* The word 'available' in the International Code of Zoological Nomenclature is equivalent to 'legitimate' in the International Code of Botanical Nomenclature.

When it is necessary to give more of a name to avoid confusion between names beginning with the same syllable, the same system is to be followed. For instance, two syllables are given together with the one or two first consonants of the third; or one of the last characteristic consonants of the name is added (Bertol. for Bertoloni, to distinguish it from Bertero; Michx. for Michaux, to distinguish it from Micheli).

Given names or accessory designations serving to distinguish two botanists of the same name are abridged in the same way (Adr. Juss. for Adrien de Jussieu, Gaertn. f. for Gaertner filius, R. Br. for Robert Brown, A. Br. for Alexander Braun, J. F. Gmelin for Johann Friedrich Gmelin, J. G. Gmelin for Johann Georg Gmelin, C. C. Gmelin for Carl Christian Gmelin, S. G. Gmelin for Samuel Gottlieb Gmelin, Müll. Arg. for Jean Müller of Aargau).

When it is a well-established custom to abridge a name in another manner, it is best to conform to it (L. for Linnaeus, DC. for de Candolle, St.-Hil. for Saint Hilaire, H.B.K. for Humboldt, Bonpland et Kunth, F. v. Muell. for Ferdinand von Mueller).

[NOTE: See also pp. 179–187 for abbreviations of fungal and lichen authors.]

Recommendation 46B

When a name has been published jointly by two authors, the names of both should be cited, linked by means of the word *et* or by an ampersand (&).

When a name has been published jointly by more than two authors, the citation should be restricted to that of the first one followed by *et al.*

[EXAMPLES: *Lecanora congesta* Clauz. & Vězda (*Portug. Acta Biol.*, B, **9**: 331, 1969) or *L. congesta* Clauz. et Vězda; *Leucosporidium antarcticum* Fell, Statzell, I. L. Hunter & Phaff (*Antonie van Leeuwenhoek* **35**: 438, 1970) should be shortened to *L. antarcticum* Fell *et al.*]

Recommendation 46C

When an author who first validly publishes a name ascribes it to another person, the correct author citation is the name of the actual publishing author, but the name of the other person, followed by the connecting word *ex*, may be inserted before the name of the publishing author, if desired. The same holds for names of garden origin ascribed to "hort." (hortulanorum).

[EXAMPLES: *Calocybe* Kühner ex Donk (*Beih. Nova Hedwigia* **5**: 42, 1962), *Volvariella murinella* (Quél.) Moser ex Dennis *et al.* (*Trans. Br. mycol. Soc.* **43** *Suppl.*: 167, 1960).]

Recommendation 46D

When a name with a description or diagnosis (or reference to a description or diagnosis) supplied by one author is published in a work by another author, the word *in* should be used to connect the names of the two authors. In such cases the name of the author who supplied the description or diagnosis is the most important and should be retained when it is desirable to abbreviate such a citation.

[EXAMPLES: *Thamnostylum* Arx & Upadhyay in Arx (*Gen. Fungi Sp. Cult.*: 247, 1970) or *Thamnostylum* Arx & Upadhyay; *Usnea crassa* Zammuto in Dodge (*Trans. Am. microsc. Soc.* **84**: 521, 1965) or *Usnea crassa* Zammuto. The word '*apud*' has sometimes been used instead of '*in*'.]

Recommendation 46E

When an author who first validly publishes a name ascribes it to an author who published the name before the starting point of the group concerned (see Art. 13), the author citation may include, when such indication is considered useful or desirable, the name of the pre-starting-point author followed by *ex* as in Rec. 46C.

[EXAMPLE: *Boletus piperatus* Bull. ex Fr. (*Syst. mycol.* **1**: 388, 1821) as *B. piperatus* Bull. (*Hist. Champ. Fr.* **1**: 318, 1791) is a pre-starting point name (Art. 13). Square brackets, i.e. *B. piperatus* [Bull.] Fr., were once recommended in the Code (Lanjouw *et al.*, 1966) but should not now be used (see, however, *Taxon* **22**: 702, 1973); the use of the form *B. piperatus* Bull. per Fr. advocated by some mycologists has never been recommended by the Code and so is to be avoided for the time being. It should be noted that pre-starting point authors are only correctly cited when the nomenclatural type of the taxon is the same as that of the pre-starting point author (see *Taxon* **6**: 245–256, 1957 and **14**: 180–184, 1965).]

Recommendation 46F

Authors of new names of taxa should not use the expression *nobis* (*nob.*) or a similar reference to themselves as an author citation but should cite their own names in each instance.

[EXAMPLE: *Amanita asperoides* Heim (*Rev. mycol.* **28**: 7, 1963) was introduced as '*Amanita asperoides* nob.'; under this recommendation '*Amanita asperoides* Heim sp. nov.' would be a preferred form. The use of 'm.' or 'mihi' is similarly to be avoided.]

Article 47

An alteration of the diagnostic characters or of the circumscription of a taxon without the exclusion of the type does not warrant the citation of the name of an author other than the one who first published its name.

[EXAMPLES: See under Art. 51.]

Recommendation 47A

When the alteration mentioned in Art. 47 has been considerable, the nature of the change may be indicated by adding such words, abbreviated where suitable, as *emendavit* (*emend.*) (followed by the name of the author responsible for the change), *mutatis characteribus* (*mut. char.*), *pro parte* (*p.p.*), *excluso genere* or *exclusis generibus* (*excl. gen.*), *exclusa specie* or *exclusis speciebus* (*excl. sp.*), *exclusa varietate* or *exclusis varietatibus* (*excl. var.*), *sensu amplo* (*s. ampl.*), *sensu stricto* (*s. str.*), etc.

[EXAMPLES: See under Art. 51.]

Article 48

When an author circumscribes a taxon in such a way as to exclude the original type of the name he uses for it, he is considered to have published a later homonym that must be ascribed solely to him.

[EXAMPLE: Batista and Ciferri (*Saccardoa* **2**: 155, 1963) specifically excluded *Naetrocymbe fuliginea* Körb. from their treatment of the genus *Naetrocymbe* as it had been shown to belong to *Coccodinum* Massal. and designated a neotype (*N. mauritiae* Bat.) for the genus. As *N. fuliginea* was the only original species in *Naetrocymbe* Körb. (*Parerg. Lich.*: 442, 1865) it is the holotype (monotype) of that genus and these authors are treated as having described a new genus *Naetrocymbe* Bat. & Cif. which is a later homonym of Körber's name; Batista and Ciferri's name is therefore illegitimate under Art. 64.]

Retention of a name in a sense that excludes the type can be effected only by conservation. When a name is conserved with a type different from that of the original author, the author of the name as conserved, with the new type, must be cited.

[EXAMPLE: '*Phoma* Fr. em.' was used by Saccardo (*Michelia* **2**: 4, 1880) for two species (*P. denigrata* Desm. and *P. herbarum* Westend.) neither of which was cited by Fries in his original description of the genus (*Syst. mycol.* **2**: 546, 1823) which is typified by *P. pustulata* Pers. ex Fr. Most later authors have used the generic name *Phoma* in

Saccardo's and not Fries' sense and *Phoma* Sacc. (with *P. herbarum* as lectotype) has consequently been conserved against *Phoma* Fr.]

Article 49

When a genus or a taxon of lower rank is altered in rank but retains its name or epithet, the author who first published this as a legitimate name or epithet (the author of the basionym) must be cited in parentheses, followed by the name of the author who effected the alteration (the author of the combination). The same holds when a taxon of lower rank than genus is transferred to another taxon, with or without alteration of rank.

[EXAMPLES: When *Strigula pulchella* Müll. Arg. is treated as a variety of *S. nemathora* Mont. it becomes *S. nemathora* var. *pulchella* (Müll. Arg.) R. Sant.; when *Camptosphaeria* Fuckel is treated as a subgenus of *Cercophora* Fuckel it becomes *Cercophora* subgen. *Camptosphaeria* (Fuckel) Lundq.; when *Ramalina* sect. *Ramalinopsis* Zahlbr. is raised to generic rank it becomes *Ramalinopsis* (Zahlbr.) Follm. & Huneck.]

Section 4. GENERAL RECOMMENDATIONS ON CITATION

Recommendation 50A

In the citation of a name published as a synonym, the words 'as synonym' or *pro syn.* should be added.

When an author has published as a synonym a manuscript name of another author, the word *ex* should be used in citations to connect the names of the two authors (see Rec. 46C).

[EXAMPLE: In the case of the example given under Rec. 34A, for Donk's name the form '*Gilbertia* Donk ex Gilbert pro syn.' is recommended for future citations of this name in taxonomic works.]

Recommendation 50B

In the citation of a *nomen nudum*, its status should be indicated by adding *nomen nudum* (*nom. nud.*).

Recommendation 50C

When a name that is illegitimate because of an earlier homonym is cited in synonymy, the citation should be followed by the name of the author of the earlier homonym preceded by the word *non*, preferably with the date of publication added. In some instances it will be advisable to cite also any later homonym, preceded by the word *nec*.

[EXAMPLE: *Lichen citrinus* Lam., *Encycl. méth. Bot.* **3**: 506 (1792) *non* Hedw., *Laub-moose* **2**: 71 (1788), *nec* Schrank, *Baiersche Fl.* **2**: 545 (1789).]

Recommendation 50D

Misidentifications should not be included in the synonymy but added after it. A misapplied name should be indicated by the words *auct. non* followed by the name of the original author and the bibliographical reference of the misidentification.

[NOTE: If the misidentification is by a single author '*sensu*' and the author misusing the name followed by '*non*' and the correct citation is often used, e.g. *Lecanora impudens sensu* P. James, *Lichenologist* **1**: 145 (1960), *non* Degel., *Svensk Bot. Tidskr.* **38**: 50 (1944). '*Auct. non*' is used in the form '*Nephroma laevigatum* auct. non Ach.'.]

Recommendation 50E

If a generic name is accepted as a *nomen conservandum* (see Art. 14 and App. III), the abbreviation *nom. cons.* should be added to the citation.

[EXAMPLES: *Venturia* Sacc. *nom. cons.*; *Xerocomus* Quél. *nom. cons.* See also example under Rec. 15A.]

Recommendation 50F

A name cited in synonymy should be spelled exactly as published by its author. If any explanatory words are required, these should be inserted in brackets. If a name is adopted with alterations from the form as originally published, it is desirable that in full citations the exact original form should be added, preferably between quotation marks.

[EXAMPLES: *Cephalosporium hypholomatis* Boedijn, *Rec. Trav. bot. néerl.* **26**: 425 (1929); as '*hypholomae*'. *Ascotricha chartarum* var. *orientalis* Castell. & Jacon., *J. Trop. Medic. Hyg.* **37**: 362 (1934); as '*Ascothrica*'. This recommendation also applies to terms denoting rank and Greek letters used to prefix infraspecific epithets. Where there are good grounds for assuming e.g. an asterisk (*) to denote a subspecies and a Greek letter (e.g., β) to denote a variety the form [subsp.] and [var.] is often used.]

Chapter V. RETENTION, CHOICE, AND REJECTION OF NAMES AND EPITHETS

Section 1. RETENTION OF NAMES OR EPITHETS OF TAXA WHICH ARE REMODELLED OR DIVIDED

Article 51

An alteration of the diagnostic characters or of the circumscription of a taxon does not warrant a change in its name, except as may be required (1) by transference of the taxon (Arts. 54–56), or (2) by its union with another taxon of the same rank (Arts. 57, 58, Rec. 57A), or (3) by a change of its rank (Art. 60).

[EXAMPLES: The genus *Ramonia* Stiz. differs somewhat from Vězda's concept of the genus (*Folia geobot. phytotax.* **1**: 155, 1966) but as the type of the genus remains in it no change in the author citation or name is required; it may be cited as *Ramonia* Stiz. or *Ramonia* Stiz. emend. Vězda (see Rec. 47A). Similarly, *Physcia tenella* (Scop.) DC. was applied in a much narrower sense than this name had been by previous authors by Bitter (*Jahrb. wiss. Bot.* **36**: 431–33, 1901) and may be cited as *P. tenella* (Scop.) DC. or *P. tenella* (Scop.) DC. emend. Bitt. but no change in the specific epithet is required.]

Article 52

When a genus is divided into two or more genera, the generic name must be retained for one of them or, if it has not been retained, must be reinstated for one of them. When a particular species was originally designated as the type, the generic name must be retained for the genus including that species. When no type has been designated, a type must be chosen.

[EXAMPLE: The genus *Agaricus* Fr. has been divided into many genera by later authors, often with the exclusion of the name *Agaricus*. As no type was designated this name has been lectotypified by *A. campestris* Fr. (*typ. cons.*) and must be retained for the genus including that species (i.e. the later name *Psalliota* (Fr.) Kumm., typified by the same species, cannot be used).]

Article 53

When a species is divided into two or more species, the specific epithet must be retained for one of them or, if it has not been retained, must be reinstated for one of them. When a particular specimen, description, or figure was originally designated as the type, the specific epithet must be retained for the species including that element. When no type has been designated, a type must be chosen.

The same rule applies to infraspecific taxa, for example, to a subspecies divided into two or more subspecies, or to a variety divided into two or more varieties.

Section 2. RETENTION OF EPITHETS OF TAXA BELOW THE RANK OF GENUS ON TRANSFERENCE TO ANOTHER GENUS OR SPECIES

Article 54

When a subdivision of a genus* is transferred to another genus or placed under another generic name for the same genus without change of rank, its epithet must be retained or, if it has not been retained, must be reinstated unless one of the following obstacles exists:

(1) The resulting combination has been previously and validly published for a subdivision of a genus based on a different type;

(2) An earlier and legitimate epithet of the same rank is available (but see Arts. 13f, 58, 59);

(3) Arts. 21 or 22 provide that another epithet be used.

[EXAMPLE: When *Parmelia* sect. *Teretiuscula* Hillm. (1936) is transferred to the genus *Alectoria* and recognized as a section it becomes *Alectoria* sect. *Teretiuscula* (Hillm.) Lamb (1964) as it is the earliest legitimate name in the same rank; the existence of *A.* sect. *Subparmelia* Degel. (1938) which is based on the same type does not affect this transfer as it is a later name in the rank of section and has to be treated as a synonym of *A.* sect. *Teretiuscula* (Hillm.) Lamb (see *Br. Antarct. Surv. Sci. Rept* **38**: 26, 1964).]

Article 55

When a species is transferred to another genus or placed under another generic name for the same genus without change of rank, the specific epithet, if legitimate, must be retained[a] or, if it has not been retained, must be reinstated[b] unless one of the following obstacles exists:

(1) The resulting binary name is a later homonym[c] (Art. 64) or a tautonym[d] (Art. 23);

(2) An earlier legitimate specific epithet is available (but see Arts. 13f, 58, 59, 72).[e]

[EXAMPLES: (a) *Puccinia pruni-spinosae* Pers. (*Syn. meth. Fung.*: 226, 1801) when transferred to the genus *Tranzschelia*, becomes *Tranzschelia pruni-spinosae* (Pers.) Dietl (*Annls mycol.* **20**: 31, 1922). (b–c) *Cephalosporium roseum* Oud. (1884), when placed in the genus *Acremonium*, must be called *Acremonium rutilum* W. Gams (1971) because of the existence of the name *Acremonium roseum* Peck (1922); if this name were transferred to another genus, however, the epithet '*roseum* Oud.' would have to be reinstated if there were no other obstacles to its employment. (d) See example under Art. 23. (e) If *Cephalosporium costantinii* F.E.Sm. (1924) is transferred to the genus *Verticillium* it must be called *V. fungicola* (Preuss) Hasselbr. (1936) as *Acrostalagmus fungicola* Preuss (1851), the basionym of Hasselbrauk's name, is the earliest available name in the rank of species.]

* Here and elsewhere in this Code the phrase 'subdivision of a genus' refers only to taxa between genus and species in rank.

When, on transference to another genus, the specific epithet has been applied erroneously in its new position to a different species, the new combination must be retained for the species to which the epithet was originally applied, and must be attributed to the author who first published it. (See Art. 7, par. 9.)

Article 56

When an infraspecific taxon is transferred without change of rank to another genus or species, the original epithet must be retained or, if it has not been retained, must be reinstated unless one of the following obstacles exists:

(1) The resulting ternary combination has been previously and validly published for an infraspecific taxon based on a different type, even if that taxon is of different rank;

(2) An earlier legitimate epithet is available (but see Arts. 13f, 58, 59, 72);

(3) Arts. 24 or 26 provide that another epithet be used.

[EXAMPLE: *Strigula complanata* var. *stellata* Nyl. & Cromb. (*Trans. Linn. Soc. Lond., ser.* 2, **2**: 114, 1883) when transferred as a variety to *S. elegans*, becomes *S. elegans* var. *stellata* (Nyl. & Cromb.) R. Sant. (*J. Ecol.* **40**: 128, 1952).]

When, on transference to another genus or species, the epithet of an infraspecific taxon has been applied erroneously in its new position to a different taxon of the same rank, the new combination must be retained for the taxon to which the original combination was applied, and must be attributed to the author who first published it (see Art. 7, par. 9).

Section 3. CHOICE OF NAMES WHEN TAXA OF THE SAME RANK ARE UNITED

Article 57

When two or more taxa of the same rank are united, the oldest legitimate name or (for taxa below the rank of genus) the oldest legitimate epithet is retained, unless a later name or epithet must be accepted under the provisions of Arts. 13f, 22, 26, 58, or 59. The author who first unites taxa bearing names or epithets of the same date has the right to choose one of them, and his choice must be followed.

[EXAMPLES: Henssen (*Symb. bot. upsal.* **18**(1): 99, 1963), when treating the genera *Polychidium* (Ach.) Gray (1821), *Leptogidium* Nyl. (1873), *Dendriscocaulon* Nyl. (1888) and *Leptodendriscum* Vain. (1890) as a single genus, correctly adopted the oldest of these generic names, *Polychidium* (Ach.) Gray, for the resulting genus. In 1826 Persoon published the three names *Polyporus corrugatus* Pers., *P. fusco-badius* Pers. and *P. scabrosus* Pers. in the same work; Fries (*Epicr.*: 469, 1838), in uniting these species under one name, adopted the name *P. scabrosus*, and this name must therefore be retained for the combined taxon by later authors.]

Recommendation 57A

Authors who have to choose between two generic names should note the following suggestions:

(1) Of two names of the same date, to prefer that which was first accompanied by the description of a species.

(2) Of two names of the same date, both accompanied by descriptions of species, to prefer that which, when the author makes his choice, includes the larger number of species.

(3) In cases of equality from these various points of view, to select the more appropriate name.

Article 58

When a taxon of recent plants, algae excepted, and a taxon of the same rank of fossil or subfossil plants are united, the correct name or epithet of the recent taxon takes precedence.

Section 4. NAMES OF FUNGI WITH A PLEOMORPHIC LIFE CYCLE AND OF FOSSILS ASSIGNED TO FORM-GENERA

Article 59

In Ascomycetes and Basidiomycetes (inclusive of Ustilaginales) with two or more states in the life cycle (except those which are lichen-fungi), the correct name of all states which are states of any one species is the earliest legitimate name typified by the perfect state. The perfect state is that which is characterized by the presence of asci in the Ascomycetes, cells of the kind giving rise to basidia in the Uredinales and in the Ustilaginales, or basidia or organs which bear basidia in the other orders of the Basidiomycetes. However, the provisions of this Article shall not be construed as preventing the use of names of imperfect states in works referring to such states; in the case of imperfect states, a name refers only to the state represented by its type.

When not already available, specific or infraspecific names for imperfect states may be proposed at the time of publication of the name for a perfect state or later, and may contain either the specific epithet applied to the perfect state or any other epithet available.

The nomenclatural type of a taxon whose name has been ascribed to a genus characterised by a perfect state must be one of which the original description or diagnosis included a description or diagnosis of the perfect state (or of which the possibility cannot be excluded that the original author included the perfect state in his description or diagnosis). If these requirements are not fulfilled the name, although validly published, shall be considered illegitimate.

The combination of the specific or infraspecific epithet of a name typified by an imperfect state with a name of a genus characterized by a perfect state shall be considered not validly published as a new combination, but shall be considered the validly published name of a new taxon if the author provides a description (in Latin, on or after 1 Jan. 1935) of the perfect state and indicates a type (on or after 1 Jan. 1958) showing the perfect state, and shall be attributed to the author of that name and to him alone. However, publication on or after 1 Jan. 1967 of a combination based on an imperfect state and applied inclusive of the perfect state shall not be considered the valid publication of a new name of the perfect state.

[EXAMPLES: The name *Ravenelia cubensis* Arth. & Johnst. (*Mem. Torrey bot. Club* **17**: 118, 1918), based on a specimen bearing only uredinia (an imperfect state), was validly published but is illegitimate because the genus *Ravenelia* is typified by a perfect state. The correct name for this fungus is *Uredo cubensis* Cummins (*Mycologia* **48**: 607, 1956), published (incorrectly) as '(Arth. & Johnst.) Cumm. comb. nov.'. Similarly the name *Gibberella lateritia* (Nees ex Link) Snyder & H. Hans. (*Am. J. Bot.* **32**: 664, 1945) is illegitimate as it is based on the name *Fusarium lateritium* Nees ex Link which is typified by an imperfect state while the genus *Gibberella* Sacc. is typified by a perfect state.

The combination '*Mycosphaerella aleuritidis* (Miyake) Ou comb. nov.' (*Sinensia* **11**: 183, 1940), based on *Cercospora aleuritidis* Miyake, is not validly published as a new combination. As Ou's 'combination' was accompanied by a Latin description of the perfect state, he is considered to have described a new species, *Mycosphaerella aleuritidis* Ou, which must be typified by material bearing the perfect state studied by Ou. Any name published in this manner after 1 January 1967 is not validly published as either a new name or a new combination (e.g. '*Mycosphaerella rhois* (Saw. & Katsuki) Chichang', *Bot. Bull. Acad. sin., Taipei, n.s.* **8**: 140, 1967, based on *Cercospora rhois* Saw. & Katsuki, was not validly published although a Latin description of the perfect state was given).

'*Corticium microsclerotia* (Matz) G. F. Web.' (*Phytopathology* **29**: 565, 1939), based on *Rhizoctonia microsclerotia* Matz, was published as a new combination with a description of the perfect state in English but is considered to be not validly published as Matz's type does not show the characteristics of a perfect state genus. This name is also not validly published as a new name as no Latin diagnosis was provided and the valid publication of the name *Corticium microsclerotia* G. F. Web. dates from 1951 (*Mycologia* **43**: 728, 1951), when a Latin diagnosis of the perfect state was supplied.

It should be noted that the case of perfect state taxa described in imperfect state genera is not discussed in the present form of this Article. At present different authors treat such names in different ways (compare e.g. *Can. J. Bot.* **50**: 2613–28, 1972, with *Stud. mycol.* **2**: 1–65, 1972) and a proposal to cover this situation is to be considered by the next International Botanical Congress in 1975 (Hawksworth and Sutton, 1974).

For further information and discussion on Art. 59 see also *Taxon* **9**: 231–241 (1960) and **11**: 70–71 (1962). The application of this Article to Persoon's names in *Uredo* is discussed in *Taxon* **17**: 179–180 (1968).]

As in the case of pleomorphic fungi, the provisions of the Code shall not be construed as preventing the use of names of form-genera in works referring to such taxa.

Section 5. CHOICE OF NAMES WHEN THE RANK OF A TAXON IS CHANGED

Article 60

When the rank of a genus or infrageneric* taxon is changed, the correct name or epithet is the earliest legitimate one available in the new rank. In no case does a name or an epithet have priority outside its own rank.

[EXAMPLES: *Nectria* subgen. *Hyphonectria* Sacc. (*Syll. Fung.* **2**: 501, 1883) was raised to generic rank in 1938 as *Hyphonectria* (Sacc.) Petch (*Trans. Br. mycol. Soc.* **21**: 270, 1938) but the correct name for the genus is *Nectriopsis* Maire (*Annls mycol.* **9**: 323, 1911), based on the same type species, as it is the earliest available legitimate name in the rank of genus. *Lichen reticulatus* Wulf. (*Collect. Bot.* **2**: 187, 1789), when treated as a

---

* Here and elsewhere in the Code the term 'infrageneric' refers to all ranks below that of genus.

form of *Alectoria pubescens* (L.) R. Howe must be called *A. pubescens* f. *subciliata* (Nyl.) D. Hawksw. (*Lichenologist* **5**: 236, 1972) as the epithet '*subciliata*' was used in the rank of forma in 1936 and is the earliest available name in the rank of form. When *Cetraria islandica* var. *crispa* Ach. (*Lich. univ.*: 513, 1810) is treated as a species it must be called *C. ericetorum* Opiz (*Seznam rostlin'květeny české*: 175, 1852) and not *C. crispa* (Ach.) Nyl. (*Bull. Soc. Linn. Normand., sér.* 4, **1**: 202, 1887).]

Recommendation 60A

(1) When a section or a subgenus is raised in rank to a genus, or the inverse change occurs, the original name or epithet should be retained unless it is contrary to this Code.

(2) When an infraspecific taxon is raised in rank to a species, or the inverse change occurs, the original epithet should be retained unless the resulting combination is contrary to this Code.

(3) When an infraspecific taxon is changed in rank with the species, the original epithet should be retained unless the resulting combination is contrary to this Code.

Article 61

When a taxon of a rank higher than genus and not higher than family is changed in rank, the stem of the name is to be retained and only the termination altered (*-inae, -eae, -oideae, -aceae*), unless the resulting name is rejected under Arts. 62–72.

[EXAMPLE: When the subfamily Theleboloideae Brumm. (*Persoonia Suppl.* **1**: 59, 1967) is raised to the rank of family it becomes the Thelebolaceae (Brumm.) Eckblad (*Nytt Mag. Bot.* **15**: 22, 1968).]

Section 6. REJECTION OF NAMES AND EPITHETS

Article 62

A legitimate name or epithet must not be rejected merely because it is inappropriate or disagreeable, or because another is preferable or better known, or because it has lost its original meaning.

[EXAMPLES: *Hypoxylon* Bull. ex Fr. and *Peltigera* Willd. cannot be changed to '*Hypoxylum*' and '*Peltophora*' as proposed by Clements (*Gen. Fung.*, 1909). Gilbert (in Bresadola, *Icon. mycol.* **27**: 63, 1940) introduced the name '*Venenaria*' to replace *Venenarius* Earle so that the endings of specific epithets in this and allied genera would be grammatically similar; this is unacceptable under this Article. *Puccinia sorghi* Schw. (1831) was stated to grow on *Sorghum* but although the type specimen proved to be on *Zea* and not *Sorghum* the name must be retained for this rust on *Zea* and neither of the names *P. maydis* Béreng. (1844) and *P. zeae* Béreng. (1845) can be applied to it merely because the epithet '*sorghi*' has lost its original meaning.]

The names of species and of subdivisions of genera assigned to genera whose names are conserved later homonyms, and which had earlier been assigned to the genera under the rejected homonymic names, are legitimate under the conserved names without change of authorship or date if there is no other obstacle under the rules.

[EXAMPLE: *Phoma herbarum* Westend. (1852) is to be accepted as legitimate although the genus *Phoma* Fr. (1821) to which it was assigned by Westendorp is rejected and the genus in which it is now placed is *Phoma* Sacc. (1880) *nom. cons.*]

Article 63

A name is illegitimate and is to be rejected if it was nomenclaturally superfluous when published, i.e. if the taxon to which it was applied, as circumscribed by its author, included the type of a name or epithet which ought to have been adopted under the rules.

The inclusion of a type (see Art. 7) is here understood to mean the citation of a type specimen, the citation of the illustration of a type specimen, the citation of the type of a name, or the citation of the name itself unless the type is at the same time excluded either explicitly or by implication.

[EXAMPLES: The generic name *Torrubia* Lév. ex Tul. (1865) when first validly published included *Cordyceps militaris* (Fr.) Link; as *C. militaris* is the type species of the genus *Cordyceps* (Fr.) Link (1833) the name *Cordyceps* should have been accepted as the name of the genus and *Torrubia* is consequently a nomenclaturally superfluous name which must be rejected. Similarly, Chenantis (1919), when introducing the generic name *Lasiosordaria* Chen., included the lectotype of *Lasiosphaeria* Ces. & de Not. and the monotypes of *Arnium* Nits. ex Fuckel and *Bombardia* (Fr.) Karst. within his genus *Lasiosordaria*; the generic name *Lasiosordaria* is consequently illegitimate three times over under this Article and must be rejected (see Lundqvist, 1972). See also the example on p. 139.]

A name is not illegitimate, even if it was nomenclaturally superfluous when published, if it is a new combination the epithet of whose basionym is legitimate. When published it is incorrect, but it may become correct later.

Article 64

A name is illegitimate and must be rejected if it is a later homonym, that is, if it is spelled exactly like a name previously and validly published for a taxon of the same rank based on a different type. Even if the earlier homonym is illegitimate, or is generally treated as a synonym on taxonomic grounds, the later homonym must be rejected, unless it has been conserved.

Note. Mere orthographic variants of the same name are treated as homonyms when they are based on different types (see Arts. 73 and 75).

[EXAMPLES: The Hyphomycete genus *Dactylina* Arnaud ex Subram. (1964) is a later homonym of *Dactylina* Nyl. (1860), a well known genus of lichens, and consequently must be rejected. The lichen genus *Squamaria* Hoffm. (1789) is a later homonym of the flowering plant genus *Squamaria* Zin. (1757) and so this is also illegitimate and must be rejected. Although *Clavaria* Vaill. ex Fr. (1821) is a later homonym of the algal genus *Clavaria* Stackh. (1816), *Clavaria* Vaill. ex Fr. has been conserved and so may be retained. *Lichen reticulatus* Nödh. (1801) and *L. reticulatus* Ach. (1798) are both later homonyms of *L. reticulatus* Wulf. (1789) and consequently both must be rejected. The fungal genus *Porotheleum* (Pers. ex Fr.) Fr. (1825) is a later homonym of *Porothelium* Eschw. (1824), a genus of lichens, as it is an orthographic variant and so must be rejected. See also under Arts. 65 and 75.]

The names of two subdivisions of the same genus, or of two infraspecific taxa within the same species, even if they are of different rank, are treated as homonyms if they have the same epithet and are not based on the same type. The same epithet may be used for subdivisions of different genera, and for infraspecific taxa within different species.

[EXAMPLES: Schade (*Ber. dt. bot. Ges.* **69**: 285, 1956) described *Cladonia sylvatica* f. *caerulescens* Schade and *C. sylvatica* f. *decumbens* subf. *caerulescens* Schade based on different types. As these epithets are employed in the same species they are treated as homonyms and Schade (*Abh. Ber. Naturk. Görlitz* **35**: 69, 1957) correctly introduced the new name *C. sylvatica* f. *decumbens* subf. *caerulea* Schade for subf. *caerulescens*. The names *Fusarium avenaceum* var. *pallens* Wollenw. and *F. lateritium* var. *pallens* Wollenw. are both acceptable as the epithet *pallens* is employed for different infraspecific taxa within different species.]

When the same new name is simultaneously published for more than one taxon, the first author who adopts it in one sense, rejecting the other, or provides another name for one of these taxa is to be followed.

Article 65

A name is illegitimate and is to be rejected if it is the name of a taxon which on transfer of that taxon from the animal to the plant kingdom becomes, at the time of such transfer, a homonym of a name for a plant taxon.

If a taxon is transferred from the plant kingdom to the animal kingdom, its name or names retain their status in botanical nomenclature for purposes of homonymy. In all other cases, the name of a plant is not to be rejected merely because it is the same as the name of an animal.

[EXAMPLES: The generic name *Empusa* Cohn (*Hedwigia* **1**: 60, 1855) was rejected by Fresenius (*Bot. Z.* **14**: 883, 1856) as it was a later homonym of *Empusa* Illiger (1798) [Orthoptera], *Empusa* Hübner (1819) and *Empusa* Hübner (1821) [different genera of Lepidoptera], and *Empusa* Lindley (1824) [Orchidaceae], and the new name *Entomophthora* Fres. proposed for Cohn's genus. Linnaeus treated generic names of animals and plants as homonyms when spelt the same and rejected one of them but this practice is no longer followed, botanical and zoological nomenclature being treated as distinct (Principle I). The existence of Hübner's and Illiger's names is consequently irrelevant and they do not have to be considered when assessing the nomenclatural status of those of Cohn and Lindley. Cohn's name is rejected simply because of Lindley's name (Art. 64) and Lindley's name is accepted and not treated as a homonym of the zoological names under the provisions of this Article. Similarly, the generic name *Drosophila* Quél. (1886) [Agaricaceae] is not treated as a later homonym of *Drosophila* Fállen (1823) [Diptera] although the name is probably familiar to most mycologists as a genus of insects not of the Agaricaeae.]

Article 66

An epithet of a subdivision of a genus is illegitimate and is to be rejected in the following special cases:

(1) If it was published in contravention of Arts. 51, 54, 57, 58 or 60, i.e. if its author did not adopt the earliest legitimate epithet available for the taxon with its particular circumscription, position, and rank.

(2) If it is an epithet of a type subgenus or section which contravenes Art. 22.

Note 1. Illegitimate epithets are not to be taken into consideration for purposes of priority (see Art. 45) except in the rejection of a later homonym (Art. 64).

Note 2. An epithet originally published as part of an illegitimate name may be adopted later for the same taxon, but in another combination (see Art. 72).

Article 67

A specific or infraspecific epithet is illegitimate and is to be rejected if it was published in contravention of Arts. 51, 53, 55, 56, or 60, i.e. if its author did not adopt the earliest legitimate epithet available for the taxon with its particular circumscription, position, and rank. Such an epithet is also illegitimate if it was published in contravention of Art. 59.

Note. The publication of a name containing an illegitimate epithet is not to be taken into consideration for purposes of priority (see Art. 45) except in the rejection of a later homonym (Art. 64).

[EXAMPLE: When Acharius (*Lich. suec. Prod.*: 64, 1798) described *Lichen parasemus* Ach. he included the legitimate name *L. limitatus* Scop. (*Fl. Carniol.*, ed. 2, **2**: 363, 1772) as a synonym. Scopoli's name should have been adopted and *L. parasemus* is consequently an illegitimate name which must be rejected and (see Art. 7) automatically typified by the type of Scopoli's name (see *Lichenologist* **1**: 158–168, 1960; *Fld Stud.* **3**: 535–578, 1972). See also examples under Art. 59.]

Article 68

A specific epithet is not illegitimate merely because it was originally published under an illegitimate generic name, but is to be taken into consideration for purposes of priority if the epithet and the corresponding combination are in other respects in accordance with the rules. In the same way an infraspecific epithet may be legitimate even if originally published under an illegitimate name of a species or infraspecific taxon.

[EXAMPLE: *Gyrophora arctica* Ach., although published under the illegitimate generic name *Gyrophora* Ach. (a superfluous name for *Umbilicaria* Hoffm., 1789, and consequently illegitimate under Art. 63), is nevertheless a legitimate name as in other respects this name was in accordance with the Code. The combination *Umbilicaria arctica* (Ach.) Nyl. is consequently also legitimate and should be attributed to '(Ach.)Nyl.' and not merely 'Nyl.'.]

Note. An illegitimate epithet may be adopted later for the same taxon, but in another combination (see Art. 72).

[NOTE: See examples cited under Art. 72.]

Article 69

A name is to be rejected if it is used in different senses and so has become a long-persistent source of error.

[EXAMPLES: The name *Erysiphe communis* Wallr. ex Fr. (1829) was applied by Fries to many of the Erysiphaceae of later authors. In 1900 Salmon treated this name as a synonym of *E. polygonia* DC. ex St.-Am. (1821) and in 1933 Blumer used it in a restricted sense but not for species on Leguminosae. As Fries' name has to be typified by *Alphitomorpha communis* α *leguminosarum* Wallr. (1819) this name has been applied in different senses and consequently should be rejected (see *Trans. Br. mycol. Soc.* **48**: 541, 1965). In some cases it may be preferable to reject a name before it has become a 'long-persistent source of error' when it is discovered that it has been applied erroneously (see *Trans. Br. mycol. Soc.* **57**: 317–324, 1971).]

Article 70

A name is to be rejected if it is based on a type consisting of two or more entirely discordant elements, unless it is possible to select one of these elements as a satisfactory type.

[EXAMPLES: The genus *Heterobasidium* Massee was based on the spores of a discomycete and the thallus of a member of the Thelephoraceae and as neither element constitutes a satisfactory type the name has to be rejected. Similarly, *Langloisula* Ellis & Everh. was based on material probably of a species of *Vararia* Karst. parasitized by a species of *Chromosporium* Corda but as the generic description was based on the characters of both elements neither can provide a satisfactory type and the name *Langloisula* must be rejected (see *Fungus* **26**: 21, 1956). At least nine different names, mainly in species rank, have been based on ascocarps of *Lecidea* species parasitized by the lichenicolous fungus *Arthonia intexta* Almq., and as their descriptions were based on both the host ascocarp and the asci and spores of the parasite they must be rejected (see *Ber. dt. bot. Ges.* **82**: 209–220, 1969).]

Article 71

A name is to be rejected if it is based on a monstrosity.

[EXAMPLES: *Tremella mycetophila* Peck was based on an abnormal outgrowth of *Collybia dryophila* Quél. and so must be rejected. Similarly, *Myriadoporus* Peck was based on a monstrous specimen of *Polyporus adustus* Willd. ex Fr. and so must also be rejected.]

Article 72

A name or epithet rejected under Arts. 63–71 is replaced by the oldest legitimate name or (in a combination) by the oldest available legitimate epithet in the rank concerned. If none exists in any rank a new name must be chosen: (a) a new name (*nomen novum*) based on the same type as the rejected name may be published, or (b) a new taxon may be described and a new name published for it. If a name or epithet is available in another rank, one of the above alternatives may be chosen or (c) a new combination, based on the name in the other rank, may be published. Similar action is to be taken when the use of an epithet is inadmissible under Arts. 21, 23, and 24.

[EXAMPLE: When *Stigmatidium dendriticum* Leight. (1875) is placed in the genus *Enterographa* it is correctly called *E. dendritica* (Leight.) P. James (1965) but if it is placed in *Arthonia* it cannot be called *A. dendritica* (Leight.) Cromb. (1876), as this name is illegitimate under Art. 64, being a later homonym of *A. dendritica* (Ach.) Duf. (1818) based on a different species, and must be called *A. atlantica* P. James (1970) (a new name based on the same type as *Stigmatidium dendriticum* Leight.).]

Note. When a new epithet is required, an author may adopt an epithet previously given to the taxon in an illegitimate name if there is no obstacle to its employment in the new position or sense; the epithet in the resultant combination is treated as new.

[EXAMPLE: *Chaetomium pusillum* Ellis & Everh. (1890) is an illegitimate name (Art. 64) being a later homonym of *C. pusillum* Fr. (1829). In transferring this species to *Ascotricha* Chivers (1915) used the citation '*A. pusilla* (Ellis & Everh.) comb. nov.' but as the basionym of Chivers' name is illegitimate he is considered as having introduced a new name attributable to him alone (i.e. *A. pusilla* Chiv.) which dates from 1915 as a legitimate name. *Chaetomium ellisianum* Sacc. & Syd. (1899) was introduced as a new name for *C. pusillum* Ellis & Everh., however, and so predates Chivers' name in species rank. The correct name of this taxon in *Ascotricha* is consequently a combination based on *C. ellisianum*, i.e. *A. ellisiana* (Sacc. & Syd.) D. Hawksw. (*Trans. Br. mycol. Soc.* **54**: 322, 1970).]

Recommendation 72A

Authors should avoid adoption of an illegitimate epithet previously published for the same taxon.

Chapter VI. ORTHOGRAPHY OF NAMES AND EPITHETS AND GENDER OF GENERIC NAMES

Section 1. ORTHOGRAPHY OF NAMES AND EPITHETS

Article 73

The original spelling of a name or epithet is to be retained, except for the correction of typographic or orthographic errors (but see Art. 14, note 7).

Note 1. The words 'original spelling' in this Article mean the spelling employed when the name was validly published. They do not refer to the use of an initial capital or small letter, this being a matter of typography (see Art. 21, Rec. 73F).

Note 2. The liberty of correcting a name is to be used with reserve, especially if the change affects the first syllable and, above all, the first letter of the name.

The consonants *w* and *y*, foreign to classical Latin, and *k*, rare in that language, are permissible in Latin plant names.

The letters *j* and *v* are to be changed to *i* and *u* respectively when they represent vowels; the reverse changes are to be made when consonants are required.

[EXAMPLE: *Verrvcaria* Web. must be changed to *Verrucaria* Web.]

Diacritic signs are not used in Latin plant names. In names (either new or old) drawn from words in which such signs appear, the signs are to be suppressed with the necessary transcription of the letters so modified; for example ä, ö, ü become respectively *ae, oe, ue*; *é, è, ê* become *e*, or sometimes *ae*; *ñ* becomes *n*; *ø* becomes *oe*; *å* becomes *ao*; the diaeresis, however, is permissible (*Cephaëlis* for *Cephaelis*).*

When changes made in orthography by earlier authors who adopt personal, geographic, or vernacular names in nomenclature are intentional latinizations, they are to be preserved.

The use of a wrong connecting vowel or vowels (or the omission of a connecting vowel) in a name or an epithet is treated as an orthographic error (see Rec. 73G).

The wrong use of the terminations *i, ii, ae, iae, anus*, or *ianus*, mentioned in Rec. 73C (a, b, d), is treated as an orthographic error.

[NOTE: In some cases it may be possible to preserve an altered spelling by conservation.]

Recommendation 73A

When a new name or epithet is to be derived from Greek, the transliteration to Latin should conform to classical usage.

The *spiritus asper* should be transcribed in Latin as the letter *h*.

---

* The diaeresis should be used where required in works in which diphthongs are not represented by special type, e.g. *Cephaëlis* in works in which there is *Arisaema*, not *Arisæma*.

Recommendation 73B

When a new name for a genus, subgenus, or section is taken from the name of a person, it should be formed in the following manner:

(a) When the name of the person ends in a vowel, the letter *a* is added (thus *Ottoa* after Otto; *Sloanea* after Sloane), except when the name ends in *a*, when *ea* is added (e.g. *Collaea* after Colla), or in *ea* (as *Correa*), when no letter is added.

(b) When the name of the person ends in a consonant, the letters *ia* are added, except when the name ends in *er*, when *a* is added (e.g. *Kernera* after Kerner). In latinized names ending in *-us*, this termination is dropped before adding the suffix (*Dillenia*).

(c) The syllables not modified by these endings retain their original spelling, unless they contain letters foreign to Latin plant names or diacritic signs (see Art. 73).

(d) Names may be accompanied by a prefix or a suffix, or be modified by anagram or abbreviation. In these cases they count as different words from the original name.

[EXAMPLES: *Hansenula* Syd. and *Pseudohansenula* Mogi; *Batistamnus* Bez. & Cavalc., *Batistia* Cif., *Batistina* Peres, *Batistinula* Arx, and *Batistospora* Bez. & Herr. (all after C. A. Batista). See also Rec. 20A (h–i).]

Recommendation 73C

When a new specific or infraspecific epithet is taken from the name of a man, it should be formed in the following manner:

(a) When the name of the person ends in a vowel, the letter *i* is added (thus *glazioui* from Glaziou, *bureaui* from Bureau, *keayi* from Keay), except when the name ends in *a*, when *e* is added (thus *balansae* from Balansa, *palhinhae* from Palhina).

(b) When the name ends in a consonant, the letters *ii* are added (*ramondii* from Ramond), except when the name ends in *-er*, when *i* is added (thus *kerneri* from Kerner).

(c) The syllables not modified by these endings retain their original spelling, unless they contain letters foreign to Latin plant names or diacritic signs (see Art. 73).

(d) When epithets taken from the name of a man have an adjectival form they are formed in a similar way (e.g. *Geranium robertianum*, *Verbena hasslerana*, *Asarum hayatanum*, *Andropogon gayanus*).

(e) The Scottish patronymic prefix, 'Mac', 'Mc' or 'M', meaning 'son of', should be spelled 'mac' and united with the rest of the name, e.g. *macfadyenii* after Macfadyen, *macgillivrayi* after MacGillivray, *macnabii* after McNab, *mackenii* after M'Ken.

(f) The Irish patronymic prefix 'O' should be united with the rest of the name or omitted, e.g. *obrienii*, *brienianus* after O'Brien, *okellyi* after O'Kelly.

(g) A prefix consisting of an article, e.g. le, la, l', les, el, il, lo, or containing an article, e.g. du, dela, des, del, della, should be united to the name, e.g. *leclercii* after Le Clerc, *dubuyssonii* after DuBuysson, *lafarinae* after La Farina, *logatoi* after Lo Gato.

(h) A prefix to a surname indicating ennoblement or canonization should be omitted, e.g. *candollei* after De Candolle, *jussieui* after de Jussieu, *hilairei* after Saint-Hilaire, *remyi* after St. Rémy; in geographical epithets, however, 'St.' is rendered as *sanctus* (m.) or *sancta* (f.) e.g. *sancti-johannis*, of St. John, *sanctae-helenae*, of St. Helena.

(i) A German or Dutch prefix when it is normally treated as part of the family name, as often happens outside its country of origin, e.g. in the United States, may be included in the epithet, e.g. *vonhausenii* after Vonhausen, *vanderhoekii* after Vanderhoek, *vanbruntiae* after Mrs. Van Brunt, but should otherwise be omitted, e.g. *iheringii* after von Ihering, *martii* after von Martius, *steenisii* after van Steenis, *strassenii* after zu Strassen, *vechtii* after van der Vecht.

If the personal name is already Latin or Greek, the appropriate Latin genitive should be used, e.g. *alexandri* from Alexander, *francisci* from Franciscus, *augusti* from Augustus, *linnaei* from Linnaeus, *hectoris* from Hector.

The same provisions apply to epithets formed from the names of women. When these have a substantival form, they are given a feminine termination (e.g. *Cypripedium hookerae, Rosa beatricis, Scabiosa olgae, Omphalodes luciliae*).

Recommendation 73D

An epithet derived from a geographical name is preferably an adjective and usually takes the termination *-ensis, -(a)nus, -inus, -ianus,* or *-icus.*

[EXAMPLES: *Fibulomyces canadensis* Jülich (from Canada), *Plicaria himalayensis* Thind & Waraitch (from Himalaya) and *Sporormia antarctica* Speg. (from Antarctica). Note that it is unwise to employ a geographical name for a new species likely to be found in very different areas in the future as the epithet will not then give an indication of the range of a species and cannot be changed (see Art. 62) even if misleading (e.g. *Alectoria americana* Mot. occurs frequently in Japan as well as in North America.]

Recommendation 73E

A new epithet should be written in conformity with the original spelling of the word or words from which it is derived and in accordance with the accepted usage of Latin and latinization (see Art. 23).

[EXAMPLE: The epithet '*sinensis*' should be used for newly described taxa in preference to '*chinensis*'. Names not formed in accordance with this Recommendation (e.g. *Plasmopara chinensis* Gorl.) are not orthographic errors and cannot be changed (Art. 62). For the Latin spellings of place names see Graesse *et al.* (1971) and Stearn (1973).]

Recommendation 73F

All specific and infraspecific epithets should be written with a small initial letter, although authors desiring to use capital initial letters may do so when the epithets are directly derived from the names of persons (whether actual or mythical), or are vernacular (or non-Latin) names, or are former generic names.

Recommendation 73G

A compound name or an epithet combining elements derived from two or more Greek or Latin words should be formed, as far as practicable, in accordance with classical usage*. This may be stated as follows:

(a) In a true compound (as distinct from pseudocompounds such as *Myos-otis, nidus-avis*) a noun or adjective in a non-final position appears as a bare stem without case-ending (*Hydro-phyllum*).

(b) Before a vowel the final vowel of this stem, if any, is normally elided (*Chrys-anthemum, mult-angulus*), with the exception of Greek *y* and *i* (*poly-anthus, Meli-osma*).

(c) Before a consonant the final vowel is normally preserved in Greek (*mono-carpus, Poly-gonum, Coryne-phorus, Meli-lotus*), except that *a* is commonly replaced by *o* (*Hemero-callis* from *hemera*); in Latin the final vowel is reduced to *i* (*multi-color, menthi-folius, salvii-folius*).

(d) If the stem ends in a consonant, a connecting vowel (*o* in Greek, *i* in Latin) is inserted before a following consonant (*Odont-o-glossum, cruc-i-formis*).

Some irregular forms, however, have been extensively used through false analogy (*atro-purpureus*, on the analogy of pseudo-compounds such as *fusco-venatus* in which *o* is the ablative case-ending). Others are used as revealing etymological distinctions (*caricae-formis* from *Carica*, as distinct from *carici-formis* from *Carex*; *tubae-florus*, with trumpet-shaped flowers, as distinct from *tubi-florus*, with tube-like or tubular flowers). Where such iregularities occur in the original spelling of existing compounds, this spelling should be retained.

Note. The hyphens in the above examples are given solely for explanatory

* See also *Taxon* 23: 163–177 (1974).

reasons. For the use of hyphens in botanical names and epithets see Arts. 20 and 23.

Recommendation 73H

Epithets of fungus names derived from the generic name of the host plant should be spelled in accordance with the accepted spelling of this name; other spellings must be regarded as orthographic variants and should be corrected.

[EXAMPLES: *Phyllachora anonicola* Chard. and *Meliola albizziae* Hansf. & Deight. must be changed to *P. annonicola* and *M. albiziae*, respectively, as the spellings *Annona* and *Albizia* are now accepted in preference to *Anona* and *Albizzia*.]

Recommendation 73I

The etymology of new names and epithets should be given when the meaning of these is not obvious.

[NOTE: This is often done by placing details of the etymology of a name in brackets or as a footnote in the place of valid publication of the name. Walker and Smith (*Trans. Br. mycol. Soc.* **58**: 461, 1972) in introducing the new species *Leptosphaeria korrae* Walk. & A.M.Sm. gave the etymology in the form '(etym. *korra*, gramen, in lingua una aboriginum Australiae Meridionalis)'.]

Article 75

When two or more generic names are so similar that they are likely to be confused,* because they are applied to related taxa or for any other reason, they are to be treated as variants, which are homonyms when they are based on different types.

[EXAMPLES: *Craterellus* Pers. (1825) and *Craterella* Gray (1821), based on different nomenclatural types, are treated as variants and so homonyms. If *Pilophorus* Th. Fr. (1857) and *Pilophoron* (Tuck.) Tuck. (1858) were treated as different genera the later name would have to be rejected as a later homonym (see *Taxon* **21**: 327–329, 1972). *Desmazieria* Mont. (1852) and *Desmazeria* Dumort. (1822) would be treated as variants if in the same group but as one is a lichen and one a grass genus both are retained as they are unlikely to be confused. *Lachnocladium* Lév. (1849) was introduced as a new name for *Eriocladus* Lév. (1846) as Léveillé considered his earlier name to be a later homonym of the orchid genus *Eriocladium* Lindley (1839); as these are not likely to be confused *Eriocladus* may be retained. See also examples under Art. 64.]

The same applies to specific epithets within the same genus and to infraspecific epithets within the same species.

[EXAMPLES: The following pairs of epithets are treated as variants which are homonyms:– *chinensis* and *sinensis; ceylanica* and *zeylanica; napaulensis, nepalensis,* and *nipalensis; polyanthemos* and *polyanthemus; macrostachys* and *macrostachyus; heteropus* and *heteropodus; poikilantha* and *poikilanthes; pteroides* and *pteroideus; trinervis* and *trinervius; macrocarpon* and *macrocarpum; trachycaulum* and *trachycaulon.* Epithets such as those in *Alectoria motykana* Bystr. (1970) and *A. motykae* D. Hawksw. (1971) are not treated as variants and so are both acceptable but authors selecting epithets for new taxa are recommended to avoid creating names which are so similar (see Rec. 23A and 23B, g–h).]

Section 2. GENDER OF GENERIC NAMES

Recommendation 75A

The gender of generic names should be determined as follows:
(1) A Greek or Latin word adopted as a generic name should retain its gender. When the gender varies the author should choose one of the alternative

---

* When it is doubtful whether names are sufficiently alike to be confused, they should be referred to the General Committee.

genders. In doubtful cases general usage should be followed. The following names, however, whose classical gender is masculine, should be treated as feminine in accordance with botanical custom: *Adonis, Diospyros, Strychnos*; so also should *Orchis* and *Stachys*, which are masculine in Greek and feminine in Latin. The name *Hemerocallis*, derived from the Latin and Greek *hemerocalles* (n.), although masculine in Linnaeus' *Species Plantarum*, should be treated as feminine in order to bring it into conformity with almost all other generic names ending in -*is*.

(2) Generic names formed from two or more Greek or Latin words should take the gender of the last. If the ending is altered, however, the gender should follow it.

Examples of names formed from Greek words:*

Modern compounds ending in -*codon*, -*myces*, -*odon*, -*panax*, -*pogon*, -*stemon*, and other masculine words should be masculine. The fact that the generic name *Andropogon* L. was originally treated as neuter by Linnaeus is immaterial.

Similarly, all modern compounds ending in -*achne*, -*chlamys*, -*daphne*, -*mecon*, -*osma* (the modern transcription of the feminine Greek word *osmé*) and other feminine words should be feminine. The fact that *Dendromecon* Benth. and *Hesperomecon* E. L. Greene were originally ascribed the neuter gender is immaterial. An exception should be made in the case of names ending in -*gaster*, which strictly speaking ought to be feminine, but which should be treated as masculine in accordance with botanical custom.

Similarly, all modern compounds ending in -*ceras*, -*dendron*, -*nema*, -*stigma*, -*stoma* and other neuter words should be neuter. The fact that Robert Brown and Bunge respectively made *Aceras* and *Xanthoceras* feminine is immaterial. An exception should be made for names ending in -*anthos* (or -*anthus*) and -*chilos* (-*chilus* or -*cheilos*), which ought to be neuter, since that is the gender of the greek words *anthos* and *cheilos*, but which have generally been treated as masculine and should have that gender assigned to them.

Examples of compound generic names where the termination of the last word is altered: *Stenocarpus, Dipterocarpus*, and all other modern compounds ending in the Greek masculine *carpos* (or *carpus*), e.g. *Hymenocarpos*, should be masculine. Those in -*carpa* or -*carpaea* however, should be feminine, e.g. *Callicarpa* and *Polycarpaea*; and those in -*carpon*, -*carpum*, or -*carpium* should be neuter, e.g. *Polycarpon, Ormocarpum*, and *Pisocarpium*.

(3) Arbitrarily formed generic names or vernacular names or adjectives used as generic names, whose gender is not apparent, should take the gender assigned to them by their authors. Where the original author has failed to indicate the gender, the next subsequent author may choose a gender, and his choice should be accepted.

[EXAMPLE: *Cordceps* Link (1833) is adjectival in form and so has no classical gender but as Link assigned to it *C. capitatus* etc., *Cordyceps* is treated as being masculine.]

(4) Generic names ending in -*oides* or -*odes* should be treated as feminine irrespective of the gender assigned to them by the original author.

When a genus is divided into two or more genera, the gender of the new generic name or names should be that of the generic name that is retained.

[EXAMPLE: When the genus *Boletus* Fr. is divided into other genera the gender of the new generic names proposed should be masculine, e.g. *Boletellus* Murr., *Xerocomus* Quél., *Gyroporus* Quel. etc. Note that names in a different gender from that of the genus from which they are split cannot have their genders changed (see the case of *Venenarius* cited under Art. 62).]

Division III. Provisions for modification of the Code

[See Stafleu *et al.* (1972, pp. 69–70).]

---

* Examples of names formed from Latin words are not given as these offer few difficulties.

Appendix II

[Lists nomina familiarum conservanda (see Stafleu *et al.*, 1972, pp. 222–238. No family names in fungi or lichens are yet conserved.]

Appendix III

[Lists nomina generica conservanda et rejicienda (see Stafleu *et al.*, 1972, pp. 239–393). 58 fungal and 32 lichen generic names are conserved and listed on pp. 254–260 and pp. 260–263, respectively, together with the rejected names. Some names in fungi and lichens are rejected in favour of names in other groups and these are listed by the conserved names on other pages. Conserved and rejected names are indicated by the generic names in Ainsworth (1971) but the list in that work does not include all names accepted for conservation for the first time by the Seattle Congress.]

GUIDE FOR THE DETERMINATION OF TYPES

[See Stafleu *et al.* (1972, pp. 75–76). The main points in this section have been included in this book on pp. 127–129 and in examples of Articles and Recommendations.]

GUIDE TO THE CITATION OF BOTANICAL LITERATURE

[See Stafleu *et al.* (1972, pp. 77–79). The main points from this section have been discussed on pp. 66–68, 114–116 and 193–199.]

VI. FUNGAL AND LICHEN AUTHORS AND THEIR HERBARIA

In order to indicate the identity of a taxon accurately it is necessary to cite the name(s) of the author(s) who published it (see Art. 46). Author's names may be abbreviated as discussed under Rec. 46A. While in many cases only the abbreviated surname is required, in others one or more initials or even a full forename (christian name) may have to be given to avoid possible confusion (e.g. A. L. Smith, A. M. Smith, A. H. Smith; Jakob Eriksson, John Eriksson). Furthermore, when abbreviating names of authors of fungi and lichens, those of botanists who have worked in other groups also have to be considered.

Suggested abbreviations for the names of some frequently cited deceased authors of fungal and lichen names are presented below, the parts not needing to be cited being placed in parentheses. I have refrained from including initials in abbreviations of some very well known mycological authors although there are botanists with the same surname who have described a few taxa (often in other groups) earlier than they did; i.e. these are treated here as 'established' in the sense of Rec. 46A. Abbreviations in this category are prefixed by an asterisk (*) below. Contracted abbreviations of author's names which are sometimes employed are indicated in square brackets but these are not recommended for general use.

For further lists of author abbreviations see Wright and Lois (1949), Stevenson (1960), Viégas (1961), Snell and Dick (1971) and Ainsworth (1971) for the fungi; Grummann (1963), Sayre *et al.* (1964) and C. F. Culberson (1969) for the lichens; Brenan (1972) for vascular plants; and Gould and Noyce (1965) for authors of genera in all plant groups. When an author is describing a new taxon it is helpful if he includes the abbreviation of his own name according to Rec. 46A.

Biographical notes on the authors of many fungal names are given by Viégas (1961) and some are included in Ainsworth (1971). The work of Barnhart (1965) is an invaluable source of reference for biographical information on both phanaerogamic and cryptogamic botanists as it includes references to obituaries, etc. and the forthcoming work of Grummann (1974) includes detailed information on lichen authors.

Mycologists engaged in monographic or revisionary studies will eventually need to trace the type material of taxa previously described in their group to ascertain which names (if any) are applicable to their 'entities' (see pp. 64–65). As type collections are often not representative of the species to which they belong they should only be borrowed in the final stages of systematic studies when the variation within species is clear. The care with which type material must be handled has already been emphasized (p. 64). Mycologists and lichenologists engaged in revisionary and monographic studies frequently encounter difficulties in tracing the type material of taxa they wish to see. To obviate this difficulty to some extent the locations of the herbaria of some of the authors listed below have been indicated by herbarium abbreviations after their dates of birth and death. Where an author's material is dispersed and the main collections are in a single institution the main herbarium is indicated first and the others are given in parentheses. It should be emphasized that type collections may be in the herbarium of the collector and not of that of the author of the taxon, and that locations of exsiccatae are not included here (see Sayre, 1969, and Stevenson, 1971, for locations of some of these). Where the location of an author's herbarium is uncertain, known herbaria (see Lanjouw and Stafleu, 1964), universities and museums in the areas he worked in should be tried; a list of the world's natural science museums is provided by Muzeelor (1971). The type material of a living author can usually be traced by writing to the author himself.

For further information on the location of herbaria see De Candolle (1880), Lanjouw and Stafleu (1954, 1957), Stafleu (1967), Commonwealth Mycological Institute (1971), Chaudri et al. (1972), J. R. Laundon (in preparation) and the forthcoming work of Grummann (1974):* some other useful references are included in Stearn (1971). A detailed account of Italian mycological herbaria has been compiled by Ciferri (1952), and Chevalier (1947) discusses damage to herbaria in World War II. These works, obituaries, and other sources have been used in compiling the data presented here but the information should not be regarded as exhaustive. The absence of any herbarium abbreviation by an author's name merely means that I have not traced the location of his herbarium and does not mean that there are no collections of his in existence. For lists of mycological culture collections see references cited on p. 36.

Herbaria are abbreviated according to the internationally accepted scheme of Lanjouw and Stafleu (1964). The work of Kent (1957) should be referred to for further details of British collections. The addresses of the herbaria cited here are listed after the list of authors and their herbaria.

Authors

Ach(arius, E. 1757–1819) H (BM, LD, PH, UPS)
Adans(on, M. 1727–1806) P
Afz(elius, A. 1750–1837) UPS
C. Ag(ardh, C. A. 1785–1859) LD
J. Ag(ardh, J. G. 1813–1901) LD
Alb(ertini, J. B. von 1769–1831)
All(escher, A. 1828–1903) HBG, M (B)
*Almq(uist, S. O. I. 1844–1923) S (H, TUR)
Amans, see St. Amans
L. Ames (L. M. 1900–1966) BPI (K)

Anders (J. 1863–1936) PR
Anzi (M. 1812–1883) TO (PAD, UPS)
Arnaud (G. 1882–1957) PC
Arnold (F. C. G. 1828–1901) M (BM, LD)
Arth(ur, J. C. 1850–1942) PUL
Atk(inson, G. F. 1854–1918) CUP
Auersw(ald, B. 1818–1870) B
C. Bab(ington, C. 1821–1889) BM (VER)
Bachm(ann, E. T. 1850–1937) B, W
Bagl(ietto, F. 1826–1916) FI, PAD, TO
Bain(ier, G. 18xx–1920)

* This work was not available when this *Handbook* went to press and information in it has consequently not been incorporated here.

Balb(is, G. B. 1765–1831) TO
Banker (H. J. 1866–1940) NY
Baroni (E. 1865–1943) FI
Barth(olomew, E. 1852–1934) FH
Bataille (F. 1850–1946)
Bat(ista, A. C. 1916–1967) URM
Batsch (A. J. G. C. 1761–1802) JE
Batt(ara, G. A. 1714–1789) PAD
J. Baumg(artner, J. 1870–19xx) W
Bausch (W. 1804–1873) B
Bay(er, A. 1882–1941) [herbarium lost]
Bayrh(offer, J. D. W. (1793–1868) FR
    [LZ destroyed]
Beauv(erie, J. J. 1874–1938)
Becc(ari, O. 1843–1920) FI (M, PAD)
*Beck (von Mannagetta und Lerchenau,
    G. R. 1856–1931) PRC
Beeli (M. 1879–19xx) BR
O. Behr (O. 1901–1957) B
Bél(anger, C. P. 1805–1881) B, G, L
Béll(ardi, C. A. L. 1741–1826) TO
    [some lost]
Beltr(amini de Casati, F. 1828–19xx)
    PAD
Berk(eley, M. J. 1803–1889) K (E)
Berl(ese, A. N. 1864–1903) PAD
Bernh(ardi, J. J. 1774–1850) JE (?US)
Bert(oloni, A. 1775–1869) BOLO
Bessey (E. A. 1877–1957) MSC, NEB
Bethel (E. 1863–1925) BPI
Bir(oli, G. 1772–1825) TO
Bisby (G. R. 1889–1958) DAOM, IMI,
    WIN
Bitt(er, F. A. G. 1873–1927) BREM, W
Bizz(ozero, G. 1852–1885) PAD
Blomb(erg, O. G. 1838–1901) LD
    (UPS)
Boedijn (K. B. 1893–1964) U (BO,
    IMI)
Boiss(ier, P. E. 1810–1885) G
Boist(el, A. B. M. 1836–1908) BM (PC)
Bolt(on, J. c. 1758–1799)
Bomm(er, E. C. 1832–1910) BR
Bond(artsev, A. S. 1877–1968)
Bonord(en, H. F. 1801–1884) ?G
Born(et, J. B. E. 1828–1911) PC
Borr(er, W. 1781–1862) K (BM)
Bory (de St.-Vincent, J. B. G. M.
    1778–1846) PC (FI)
Borzi (A. 1852–1921) B, FI
Bosch (R. J. van den 1810–1862) L
Boud(ier, J. L. É. 1828–1920) PC
B(ouly) de Lesd(ain, M. 1869–1965)
    [post-1940 with Dr G. Clauzade
    (St Joseph); most pre-1940 destroyed
    but some in GL, O, UPS, US]
H. Bourd(ot, Abbé H. 1861–1937) PC
Brandt (T. 1877–1939)
Branth (J. S. D. 1831–1917) C
Braun (A. C. H. 1805–1877) STR
    [B destroyed]

Bréb(isson, L. A. de 1798–1872) CN,
    PC
Bref(eld, O. 1839–1925) B
Bres(adola, Abbé G. 1847–1929) S
    (BPI, L, TO)
Briosi (G. 1846–1919) G, PAD, PAV
Britzelm(ayer, M. 1839–1909)
Brond(eau, L. de 1794–1859) TL
Brongn(iart, A. T. 1801–1876) PC
Broome (C. E. 1812–1886) K
R. Br(own, R. 1773–1858) BM
Bubák (F. 1865–1925) BPI
Buller (A. H. R. 1874–1944) WIN
Bull(iard, J. B. F. [P.] 1752–1793)
    [no herbarium kept]
Burl(ingham, G. S. 1872–1952) NY
Burt (E. A. 1859–1939) BPI, FH
Butl(er, E. J. 1874–1943) HCIO
Calk(ins, W. W. 1842–1914) BPI, NY,
    US
Candolle, see De Candolle
Car(estia, A. 1825–1908) PAD (FI, TO)
Carroll (I. 1828–1880) BM (CRK,
    DBN)
Cast(agne, J. L. M. 1785–1858) CN,
    G, PC
Cav(ara, F. 1857–1929) PAD
Ceng(ia-) Samb(o, M. 1888–1939)
    TOM (FI)
Ces(ati, V. 1806–1883) RO (FI, PAD)
Chard(on, G. E. 1897–1965) BPI,
    RPPR
Chaub(ard, L. A. 1785–1854)
Chenant(ais, J. 1854–1942) NTM
Chev(allier, F. F. 1796–1840) STR
    [some drawings in P]
Chod(at, R. H. 1865–1934) G
Chupp (C. D. 1886–1967) CUP (FH)
Cienk(owski, L. de 1882–1887)
Cif(erri, R. 1897–1964) BPI (PAV)
*Clem(ents, F. E. 1874–1945) NEB
R. Clem(ente, S. de R. 1777–1827) MA
Clint(on, G. P. 1867–1937) CUP, IAC,
    NY
Coem(ans, H. E. L. G. 1825–1871) BR
Cohn (F. J. 1828–1898)
Coker (W. C. 1872–1953) NCU
Col(enso, W. 1811–1899) WELT
Comm(ons, A. 1829–1919) CM (PH)
Cooke (M. C. 1825–1914) [Cke] K
    (E, PAV, PC)
Corda (A. K. J. 1809–1849) [Cda] PR
    (K)
Cornu (M. M. 1843–1901) ?PC
    [some lost]
Coss(on, E. St.-C. 1819–1889) PC
Cost(antin, J. N. 1857–1936)
Cretz(oiu, P. 1909–1946) BUCA
Cromb(ie, J. M. 1831–1906) BM
Crouan (P. L. 1798–1871) Concarneau
Croz(als, A. de 1861–1932) PC (US)

Cub(oni, G. 1852–1920) PAD
Cumm(ings, C. E. 1855–1906) MICH
G. Cunn(ingham, G. H. 1892–1962)
　IMI, K, PDD
Curr(ey, F. 1819–1881) K
*Curt(is, M. A. 1808–1872) FH (BRU,
　K, NEB, NYS)
Curzi (M. 1898–1944)
Dalla Torre (K. W. 1850–1928) [D.T.]
　IB
Dang(eard, P. C. A. 1862–1947) PC
Darb(ishire, O. V. 1870–1934) BM,
　BRIST
J. Davis (J. J. 1852–1937) WIS (BPI)
Deak(in, R. 1808–1873) BM
DC. (De Candolle, A. P. 1778–1841) G
Dearn(ess, J. 1852–1954) DAOM
　(BPI, CAN, CUP, IAC, NY)
d(e) Bary (A. 1831–1888) [d By.] BM,
　STR
Deichmann Branth, see Branth
Delacr(oix, E. G. 1858–1907)
Del(ise, D F. 1780–1841) CN
　(BM, H, STR, TL, UPS)
De Not(aris, G. 1805–1877) [DNtrs]
　RO (BM, GE, PAD, PC, TO)
Desf(ontaines, R. L. c. 1750–1833) FI
　(G, K, MPU)
Desm(azières, J. B. H. J. 1786–1862)
　BR, PC
Despr(éaux, J. M. 1794–1843) CN
Desv(aux, N. A. 1784–1856) PC
de Toni (G. B. 1864–1924) PAV (PAD)
De Wild(eman, E. A. J. [H.] 1866–1947)
Dicks(on, J. 1738–1822) BM
Died(icke, H. 1865–1940) B, WRSL
Diet(el, P, 1860–1947) B, K, S
D. Dietr(ich, D. N. F. 1799–1888)
　B, JE, L, W
Dill(enius, J. J. 1684–1747) OXF
Doidge (E. M. 1887–1965) PRE
Donk (M. A. 1908–1972) L (BO)
Duby (J. E. 1798–1885) STR (UPS)
Duf(our, J.-M. L. 1780–1865) NTM
Dumort(ier, B. C. J. 1797–1878) BR
Dunal (M. F. 1789–1856) MPU (CN)
E. Dur(and E. J. 1870–1922) CUP
Du Rietz (G. E. 1895–1967) [DR.]
　UPSV (LD, S, UPS)
Durieu (de Maisonneuve, M. C.
　1796–1878) PC
Duval (C. J. 1751–1828)
Earle (F. S. 1856–1929) NY
Eckf(eldt, J. W. 1851–1933) PH
Eggerth (C. 1860–1888)
Ehrenb(erg, C. G. 1795–1876) B (L,
　STR, UPS)
Ehrh(art, F. 1742–1795) GOET
　(LINN, MW)
Eidam (M. E. E. 1845–1901)
Eitn(er, E. 18xx–1921) WRSL (B, W)

Elenk(in, A. A. 1873–1942) LE
*Ellis (J. B. 1829–1905) NY (FH)
Endl(icher, S. L. 1804–1849) W
Erichs(en, C. F. E. 1867–1945) HBG
Eschw(eiler, F. G. 1796–1831) BR, M
A. Evans (A. W. 1868–1959) US
Everh(art, B. M. 1818–1904) MAINE,
　NY
Fabre (J. H. C. 1823–1915)
　L'Harmas
Fairm(an, C. E. 1856–1934) CUP
R. Falck (R. 1868–1955)
Farl(ow, W. G. 1844–1919) FH
Faull (J. H. 1870–19xx) BPI, FH
Faurie (U. J. 1847–1915) PC (BM, LD,
　UPS, W, WU)
Fayod (V. 1860–1900) G
Fée (A. L. A. 1789–1874) BM, FI,
　PC, STR
Feltg(en, J. 1833–1904) BR, LUX
Ferd(inandsen, C. C. F. 1879–1944) CP
Ferr(aris, T. 1874–1943) PAD
Fing(erhuth, C. A. 1798–1876)
Fink (B. 1861–1927) MICH, MINN,
　US
A. Fisch(er, A. 1858–1913)
E. Fisch(er, E. 1861–1939) BERN
　(B, BAS, KIEL, PC)
Fitzp(atrick, H. M. 1886–1949) CUP
　(FH, IAC, NY)
Flag(ey, C. 1834–1898) AL (PC)
Flah(ault, C. H. M. 1852–1935) PC
Flörke (H. G. 1764–1835) (FH)
　[B and ROST destroyed]
*Flot(ow, J. C. G. U. G. G. A. E. F.
　von 1788–1856) WRSL (L)
Forss(ell, K. B. J. 1856–1898) UPS
Fragoso, see González Fragoso
Fres(enius, J. B. G. W. 1808–1866) FR
　[some destroyed]
*Fr(ies, E. M. 1794–1878) UPS (B, LD)
Th. Fr(ies, T. M. 1832–1913) UPS (LD)
Fuckel (K. W. G. L. 1821–1876)
　[Fckl] G
Funck (H. C. 1771–1839) B
Gäum(ann, E. A. 1893–1963) BERN
Galløe (O. 1881–1965) C
Garov(aglio, S. 1805–1882) PAV (RO)
Gelt(ing, P. 1905–1964) C
Gilkey (H. M. 1886–1972) OSC
Gill(et, C. C. 1806–1896) PC
Gilman (J. C. 1890–1966)
J. Gmel(in, J. F. 1748–1804)
Gonz(ález) Frag(oso, R. 1862–1928)
W. Gord(on, W. L. 1901–1963) ATCC
Graeve (P. H. F. 1819–1866) UPS (LD)
Gray (S. F. 1766–1828) [no herbarium
　kept]
H. Greene (H. C. 1904–1967) WIS
　(BPI)
Grev(ille, R. K. 1794–1866) E (GL)

Griffiths (D. 1867–1935) BPI (NY)
Griffon (É. 1869–1912)
Groenh(art, P. 1894–1965) L (BO)
W. Grove (W. B. 1848–1938) K
Groves (J. W. 1906–1970) DAOM
Grumm(ann, V. J. 1899–1967) B
Guillaum(ot, M. Abbé 1884–1970) PC
Guillierm(ond, A. 1876–1945)
Gunn(erus, J. E. 1718–1773) TRH
Gyeln(ik, V. K. 1906–1945) BP
Haller (A. von 1708–1777) GOET, P
Halst(ed, B. D. 1852–1918)
Hampe (G. E. L. 1795–1880) BM, G
Hans(en, E. C. 1842–1909) C (K)
Hansf(ord, C. G. 1900–1966) EA
    (IMI, K)
Har(iot, P. A. 1854–1917) PC
Harkn(ess, H. W. 1821–1901) CAS (DS)
Harm(and, J. Abbé 1844–1915) DUKE
Harr(is, C. W. 1849–1910) ?NY
Harz (C. O. 1842–1906) M
Hasse (H. E. 1836–1915) NY (FH)
Hasselr(ot, T. E. 1903–1970) S
Hav(ås, J. J. 1864–1956) BG, DUKE
Häyr(én, E. 1878–1957) H
Hazsl(insky de Hazslin, F. A. 1818–1896)
    BP
Hedgc(ock, G. G. 1863–1946) BPI
Hedl(und, J. T. 1861–1953)
Hedw(ig, J. 1730–1799) G
Hegetschw(eiler-Bodmer, J. 1789–1839)
    BERN, Z
Hellb(om, P. J. 1827–1903)
P. Henn(ings, P. C. 1841–1908) B
    (HBG, K, KIEL, L, S, W)
Henriq(ues, J. A. 1838–1928) COI
Hepp (P. 1799–1867) BM, STR
Herre (A. W. C. T. 1868–1962) F
    (LAM, LD, UPS, W)
Heufl(er zu Rasen, L. S. J. D. A. von
    H. 1817–1885)
Heug(el, C. A. 1802–1876)
Hilitz(er, A. 1899–1940) PR
Hillm(an, J. 1881–1943) [destroyed
    in 1945]
Hirats(uka, N. 1873–1946) PUR
Hochst(etter, C. F. 1787–1860)
Hoffm(ann, G. F. 1760–1826) MW
    (GOET, ?LE)
H. Hoffm(ann, H. C. H. 1819–1891)
    PC
Höhn(el, F. X. R. von R. 1852–1920)
    FH (K)
Hollós (L. 1859–1940) BP
Holw(ay, E. W. D. 1853–1923) MIN,
    PUR
Hook(er) fil.(J. D. 1817–1911) BM
Hook(er, W. J. 1785–1865) BM (K)
Hornem(an, J. W. 1770–1841) C
Hoss(éus, C. C. 1878–1950) BP, TUR
House (H. D. 1878–1949) NYS

R. Howe (R. H. 1875–1932) NY
Huds(on, W. 1730–1793) [main
    herbarium destroyed in 1783;
    remains in BM, LINN]
Hue (A. M. Abbé 1840–1917) PC
    (NTM)
Hulting (J. 1842–1929) GB
Humb(oldt, F. H. A. von 1769–1859)
    P, PC (B)
Hy (F. C. Abbé 1853–1918) COLO
    (DUKE)
S. Ito (S. 1883–1962)
Jaap (O. 1864–1922) HBG
H. Jacks(on, H. S. 1883–1951) DAOM,
    DUKE, TRTC
Jacq(uin, N. J. von 1727–1817) BM
    (W)
Jacz(ewski, A. L. A. de 1863–1932) LE
Jahn (E. 1871–1942) B
Jatta (A. 1852–1912) NAP (FI)
Juel (H. O. 1863–1931) UPS (BERN,
    C, H, LD, S, W)
Junghuhn (F. W. 1812–1864) L
Kalchbr(enner, K. 1807–1886) B, BP,
    E, SAM, UPS
Kanouse (B. B. 1889–1969) MICH
P. Karst(en, P. A. 1834–1917) H
    (BPI, UPS)
C. Kauffm(an, C. H. 1869–1931) MICH
    (NY)
Keissl(er, K. von 1872–1965) W (C,
    LD)
Kernst(ock, E. 1852–1900) W
J. Kickx (J. 1803–1864) GENT (BR)
Killerm(ann, S. 1870–1956) M
E. Kirchn(er, E. O. O. von 1851–1925)
Kirschst(ein, W. 1863–1946) B (H)
Kleb(ahn, H. 1859–1942) B
Kløcker (A. 1862–1923) C
Klotz(sch, J. F. 1805–1860) B [For
    locations of Herb. Mycol. Exs. see
    Kohlmeyer (1962a).]
Köfarago-Gyelnik, see Gyelnik
Koord(ers, S. H. 1863–1919) BO
Körb(er, G. W. 1817–1885) L (G, W,
    WRSL)
Körn(icke, F. A. 1828–1908) B, BONN,
    G
Kov(ár, N. F. 1863–1925) H
Kremp(elhuber, A. von 1813–1882) M
L. Krieger (L. C. C. 1873–1940) MICH
Kühn (J. G. 1825–1910)
Kumm(er, P. 1834–1912)
Kuntze (C. E. O. 1843–1907) NY
Kunze (G. 1793–1851) L, UPS [LZ
    destroyed]
Kusano (S. 1874–1962) B, NY
Labill(ardière, J. J. H. de 1755–1834)
    FI
Lagerh(eim, N. G. 1860–1926) B, PC
Lahm (J. G. F. X. 1811–1888) B

Lam(arck, J. B. A. P. M. de 1744–1829) PC (G)
Lambotte (J. B. E. 1832–1905)
Lamy (de la Chapelle, P. M. É. 1803–1886) H, M, PC
Lång (K. G. W. 1875–1912) H, HSI
J. Lange (J. E. 1864–1941) C
Langer(on, M. C. P. 1874–1950) PC
Larb(alestier, C. D. 1838–1911) BM (CRK, FI)
Lasch (W. G. 1787–1853)
Laur(er, J. F. 1798–1873) M (BM) [B? destroyed]
Lázaro(é Ibiza, B. 1858–1921) MAF
Lebert (H. 1813–1878)
Leight(on, W. A. 1805–1889) BM (PC, UPS)
Lendn(er, A. 1873–1948)
Lett(au, G. 1878–1951) B
Lév(eillé, J. H. 1796–1870) K (E, G, L, PC) [CN destroyed 1870–71]
Leyss(er, F. W. von 1731–1815) LINN
Lib(ert, M. A. 1782–1865) BR, PC
Lightf(oot, J. 1735–1788) K [lichens lost]
Lind (J. W. A. 1874–1939) C
Lindau (G. 1866–1923) B (C, L) [some in B destroyed]
Linder (D. H. 1899–1946) FH (BPI, NY)
Lindn(er, P. 1861–1945)
Lindroth, see Liro
Linds(ay, W. L. 1829–1880) E (BM)
Link (J. H. F. 1769–1851) [Lk] B [some destroyed] (L)
Linkola (K. 1888–1942) H
L(innaeus, C. 1707–1778) LINN (S)
Linn(aeus) fil. (C. fil. 1741–1783) LINN
Liro (J. I. 1872–1943) H (IMI)
List(er, A. 1830–1908) BM
G. List(er, G. 1860–1949) BM
Litsch(auer, V. 1879–1939) W (PR)
*Lloyd (C. G. 1859–1926) BPI (US)
Lohwag (K. 1913–1970) W
Lojka (H. 1844–1887) B, BM, BP, BOL, LD, SAM, W, ZT
Long (W. H. 1867–1947) BPI
Lönnr(oth, K. J. 1826–1885) UPS
Lundell (S. 1892–1923) UPS
Lynge (B. A. 1884–1942) O (BG, BM, C, S)
McAlp(ine, D. 1848–1932) Burnley
Macbr(ide, T. H. 1848–1934) IA
Mac(oun, J. 1831–1920) CANL
Magnus (P. W. 1844–1914) HBG (?B)
Magnusson (A. H. 1885–1964) [H. Magn.] UPS
Maheu (J. M. A. 1873–1937) PC
Mains (E. B. 1890–1968) MICH
Maire (R. C. J. E. 1878–1949) AL (MPU)

Malbr(anche, A. F. 1818–1888)
Malme (G. O. A. 1864–1937) S
Mann (W. B. 1799–1839)
Marchal (Élie, 1839–1923) BR
Marchand (L. 1807–1843)
G. W. Martin (G. W. 1886–1971) IA
Mart(ius, C. F. P. von 1794–1868) AWH, BR, M
Massal(ongo, A. B. 1824–1860) VER (M, PAD, W)
C. Massal(ongo, C. B. 1852–1928) VER (PAD)
Massee (G. E. 1850–1917) K, NY
Matr(uchot, A. L. P. 1863–1921)
Mattir(olo, O. 1856–1947) TO (PAD)
Maubl(anc, A. 1880–1958)
Mérat (F. V. 1780–1851) BR, G
Mereschk(owsky, K. S. 1854–1920) BM, G
Merr(ill, G. K. 1864–1927) FH
Metzl(er, J. A. 1812–1883)
Meyen (F. J. F. 1804–1840) KIEL [B destroyed]
G. Meyer (G. F. W. 1782–1856) GOET, LE
Michaux (A. 1746–1802) [Michx.] PC (P)
Mich(eli, P. A. 1679–1737) FI
Mig(ula, W. 1863–1938) [B destroyed]
Miles (L. E. 1890–1941) ILL
J. Miller (J. H. 1890–1961) GA
Minks (A. 1846–1908)
Miyabe (K. 1860–1950)
Miyoshi (M. 1861–1939)
Moesz (G. von 1873–1946)
Möll(er, F. A. G. J. 1860–1922)
Mong(uillon, E.-L.-H. 1865–1940)
Mont(agne, J. P. F. C. 1784–1866) PC (BM, L, UPS)
Morg(an, A. P. 1836–1907) IA (FH)
Moug(eot, J. B. 1776–1858)
Mouton (V. 18xx–19xx) BR
Mudd (W. 1830–1879) BM
Mühl(enberg, G. H. E. 1753–1815)
Müll(er) Arg(oviensis, J. 1828–1896) G (BM)
O. Müll(er, O. F. 1730–1784)
Mundk(ur, B. B. 1896–1952)
Murrill (W. A. 1869–1957) NY
Nann(izzi, A. 1877–1961) SIENA
Naum(ov, N. A. 1888–1959)
Neck(er, N. J. de 1730–1793)
Nees (von Esenbeck, C. G. D. 1776–1858) STR (L, UPS)
T. Nees (von Esenbeck, T. F. L. 1787–1837) [BONN destroyed]
Nestl(er, C. G. 1778–1832) MPU
Niessl (von Mayendorf, G. 1839–1919) M
Nieuwl(and, J. [A.] A. 1878–1936)
Nits(chke, T. R. J. 1834–1883) [Nke] B [Diaporthe in DAOM]

Nöhd(en, H. A. 1775–1804)
Norm(an, J. M. 1823–1903) O (BG, TRH)
Norrl(in, J. P. 1842–1917) H
Notaris, see de Notaris
Nowak(owski, L. 1847–19xx)
*Nyl(ander, W. 1822–1899) H (BM, NY, STR, UPS)
Oakes (W. 1799–1848)
Oed(er, G. C. von E. 1728–1791)
Ohl(ert, A. O. L. 1816–1875) Kaliningrad
*Oliv(ier, P. H. 1849–1923)
Opat(owski, W. 1810–1838)
Opiz (P. M. 1787–1858) PR (PRC)
Ørst(ed, A. S. 1816–1872) C
C. Orton (C. R. 1885–1955) PUR
Otth (G. H. 1806–1874) BERN
Oudem(ans, C. A. J. A. 1825–1906) L
Overh(olts, L. O. 1890–1946) PAC
Pass(erini, G. 1816–1893) PARMA
Pat(ouillard, N. T. 1854–1926) FH (PC)
Patt(erson, F. 1847–1928) BPI
Paul(et, J. J. 1740–1826)
Paz(schke, F. O. 1843–1922) B
Peck (C. H. 1833–1917) [Pk] NYS
Penz(ig, A. G. O. 1856–1929) PAD (GE)
Pers(oon, C. H. 1761–1836) L (G, GOET, PC, STR)
Petch (P. T. 1870–1948) K
Pethybr(idge, G. H. 1871–1948)
Petr(ak, F. 1886–1973) W (S)
Petri (L. 1875–1946) FI
Peyrit(sch, J. J. 1835–1889)
Phill(ips, W. 1822–1905) K
Pit(ard, J. C. M. 1873–1927)
Plitt (C. C. 1869–1933) BPI, US
Plowr(ight, C. B. 1849–1910) K
Poetsch (I. S. 1823–1884)
Pollich (J. A. 1740–1780)
Poteb(nja, A. A. 1870–1919)
Preuss (C. G. T. 1795–1855) B
Prill(ieux, É. E. 1829–1915)
Pringsh(eim, N. 1823–1894)
Quél(et, L. 1832–1899)
Rabenh(orst, G. L. 1806–1881) B [For locations and information on Herb. Mycol. Exs. see Kohlmeyer (1962a) and Stevenson (1967).]
Racib(orski, M. 1863–1917) KRAM (ZT)
Raf(inesque-Schmaltz, C. S. 1783–1840) WIS (NY)
Ramond (L. F. E. de C. 1753–1827)
Räs(änen, V. J. P. B. 1888–1953) H
Rav(enel, H. W. 1814–1887) FH, K (BM, US)
Rea (C. 1861–1946) K
Rebent(isch, J. F. 1772–1810)
Reding(er, K. 1907–1940) W, WU

Rehm (H. 1828–1916) S (B)
Reichardt (H. W. 1835–1885)
Reinke (J. 1849–1931)
Retz(ius, A. J. 1742–1821) LD
Rick (J. E. 1869–1946) PACA
Ricker (P. L. 1878–19xx) WIS
Riddle (L. W. 1880–1921) FH
Rieb(er, X. 1860–1906)
Riess (H. 1809–1878)
Riv(olta, S. 1832–1893)
Robb(ins, C. A. S. 1874–1930) US
Roberge (M. R. 1xxx–1864) PC
Rodig (F. W. 1770–1844)
Röhl(ing, J. C. 1757–1813)
Rolland (L. 1841–1912)
Romell (L. 1854–1927) S
Rostaf(iński, J. T. 1850–1928)
Rostk(ovius, F. W. G. 1770–1848)
Rostr(up, F. G. E. 1831–1907) C, CP
Roth (A. W. 1757–1834)
Roum(eguère, C. 1828–1892) PC
Rouss(eau, M. H. 18xx–1926) BR
Roussel (H. F. A. de 1748–1812)
Roze (E. 1833–1900)
J. Russell (J. L. 1808–1873)
Rustr(öm, C. B. 1758–1826)
Rydb(erg, P. A. 1860–1931)
Sabour(aud, R. J. 1864–1938)
D. Sacc(ardo, D. 1872–1952) PAD
Sacc(ardo, P. A. 1845–1920) PAD
St.-Am(ans, J. F. B. de, 1748–1831)
E. Salm(on, E. S. 1871–1959)
Samp(aio, G. A. da S. F. 1865–1937) PO
Sandst(ede, L. E. 1859–1951)
Sarnth(ein, L. von G. 1861–1914) IB
Savicz (V. P. 1885–1972) LE
Săvul(escu, T. 1889–1963) BUCA
Saw(ada, K. 1888–1950) TAI
Schaer(er, L. E. 1785–1853) G
B. Schenk (B. 1833–1893)
Scherb(ius, J. 1769–1813)
Scherff(el, A. 1865–1939)
Schlecht(endal, D. F. L. von 1794–1866) HAL [B destroyed]
Schleich(er, J. C. 1768–1834) LAU (CANL, H)
A. Schneid(er, A. 1863–1928) ?US
Schrad(er, H. A. 1767–1836) GOET, LE
Schrank (F. von P. von 1747–1835) M
Schreb(er, J. C. D. von 1739–1810) M
Schröt(er, J. 1837–1894) WRSL
Schultz (C. F. 1765–1837)
Schulz(er von Müggenberg, S. 1802–1892)
Schum(acher, H. C. F. 1757–1830)
*Schw(einitz, L. D. von 1780–1834) ? H (BPI, FH)
Scop(oli, J. A. 1723–1788) [?M but probably lost]

Seaver (F. J. 1877–1970) NY
Secr(etan, L. 1758–1839)
Serv(ít, M. 1886–1959) BP, PRC
Setch(ell, W. A. 1864–1943) UC
Seym(our, A. B. 1859–1933) WIS
Shear (C. L. 1865–1956) BPI
A. L. Sm(ith, A. L. 1854–1937) BM
　　(K)
E. F. Sm(ith, E. F. 1854–1927)
G. Sm(ith, G. 1895–1967) IMI
Sm(ith, J. E. 1759–1828) LINN (BM)
W. G. Sm(ith, W. G. 1835–1917)
Sommerf(elt, S. C. 1794–1838) O
Sor(auer, P. C. M. 1839–1916)
Sow(erby, J. 1757–1822) BM, K (LINN)
Speg(azzini, C. L. 1858–1926) LPS
Sprague (R. 1901–1962) FH, WSP
Spreng(el, K. P. J. 1766–1833) BAS,
　　NY
Starb(äck, K. 1863–1931)
B. Stein (B. 1847–1899) WRSL
J. Stein(er, J. 1844–1918) W, WU
Stenhamm(ar, C. 1783–1866) S
Stenholm (C. 1862–1939) GB
Steud(el, E. G. von 1783–1856)
F. Stev(ens, F. L. 1871–1934) BISH,
　　ILL, NY
Stirt(on, J. 1833–1917) GLAM (BM,
　　E)
Stiz(enberger, E. 1827–1895) ZT
Stratt(on, R. 1883–19xx) OS [most lost]
Sturm (J. 1771–1848)
Suza (J. 1890–1951) PRC
Sw(artz, O. P. 1760–1818) S (UPS)
Syd(ow, P. 1851–1925) S (DAR)
H. Syd(ow, H. 1879–1946)
Szat(ala, Ö. 1889–1958) BP
Tand(on, R. N. 18xx–19xx) HICO
　　(IMI)
Tassi (F. 1851–19xx) SIENA (PAD)
Tav(ares, C. N. 1914–1972) LISU
Tayl(or, T. 1775–1848) FH (BM)
Tehon (L. R. 1895–1954) ILLS
Thaxt(er, R. 1858–1932) FH
Theiss(en, F. 1877–1919) W (FH)
Thom (C. 1872–1956) ATCC
Thüm(en, F. K. A. E. J. von 1839–
　　1892) ?PAD
Thunb(erg, C. P. P. 1743–1828) UPS
Thw(aites, G. H. K. 1811–1882) K
　　(BM)
Timkó (G. 1876–1945) BP
Tobl(er, F. 1879–1957)
Tode (H. J. 1733–1797) [herbarium
　　destroyed]
Torss(ell, G. 1811–1849)
Tourn(efort, J. P. de 1656–1708)
　　P (OXF)
Tracy (S. M. 1847–1920) BPI

Trail (J. W. H. 1851–1919) ABD
Tranz(schel, W. A. 1868–1942)
Trav(erso, G. B. 1878–1955) PAD,
　　PAV
*Trev(isan de St.-Léon, V. B. A.
　　1818–1897) PAD
Trog (J. G. 1781–1865)
Tuck(erman, E. 1817–1886) FH (US)
**C. Tul(asne, C. 1816–1884) PC
**Tul(asne, E. L. R. 1815–1885) PC
Turc(oni, M. 1879–1929) PAV
Turn(er, D. 1775–1858) BM (K, LINN)
Ulbr(ich, E. 1879–1952) B
L. Underw(ood, L. M. 1853–1907)
Unger (F. J. A. N. 1800–1870) W
Vahl (M. 1749–1804) C
Vaill(ant, S. 1669–1722)
Vain(io, E. A. 1853–1929) TUR (BM,
　　BR, C, STE, US)
v(an) Beyma (thoe Kingma, J. F. H.
　　1885–1966) CBS
v(an) Overeem (C. 1893–1927) AMD
v(an) Tiegh(em, P. E. L. 1839–1914)
Velen(ovský, J. 1858–1949) PR
Viala (P. 1859–1936)
Vill(ar[s], D. 1745–1814) GRM
Vitt(adini, C. 1800–1865) PAD
Vogl(ino, P. 1864–1933) PAD
Vuill(emin, J. P. 1861–1932) PAD,
　　PAV
Wahlenb(erg, G. 1780–1851) UPS
Wainio, see Vainio
Wakef(ield, E. M. 1886–1972) K
L. Walker (L. B. 1878–19xx) NEB
Wallr(oth, C. F. W. 1792–1857) STR
　　(PR) [B destroyed]
W. Wats(on, W. 1872–1960) BM
F. Web(er, F. 1781–1823)
Web(er, G. H. 1752–1828)
Wedd(ell, H. A. 1819–1877) PC (BM)
Weigel (C. E. von 1748–1831)
Weiss (F. W. 1744–1826)
Welw(itsch, F. 1806–1872) BM, LISU
　　(H, STE, TUR, W)
Westend(orp, G. D. 1813–1868) BR
Westerd(ijk, J. 1883–1961)
Westr(ing, J. P. 1753–1833)
Wettst(ein von Westersheim, R. von
　　1863–1931) WU
Whetz(el, H. H. 1877–1944) CUP
Wigg(ers, F. H. 1746–1811)
Willd(enow, C. L. 1765–1812) B
Will(ey, H. 1824–1907) US
G. Wils(on, G. W. 1877–19xx) NY
Wint(er, H. G. 1848–1887) B
With(ering, W. 1741–1799) BM
　　(LINN)
Wollenw(eber, H. W. 1879–1949) BSB
Worm(ald, H. 1879–1955)

**'Tul.' often used for taxa by 'Tul. & C. Tul.'

Woron(in, M. S. 1838–1903)
Wulf(en, F. X. von 1728–1805) M (W)
Yam(amoto, W. 18xx–19xx) IMI
  (TNS)
Yas(uda, A. 1868–1924)
*Zahlbr(uckner, A. 1860–1938) W
  (PAD, STE, US)
Zeller (S. M. 1885–1948) ORE (NY,
  UC)

A. W. Zimm(ermann, A. W. P.
  1860–1931)
Zoll(inger, H. 1818–1859) BR (L)
Zopf (F. W. 1846–1909) B [some
  destroyed]
Zsch(acke, H. 1867–1937)
Zuk(al, H. 1845–1900)
Zundel (G. L. I. 1885–1950) BPI
Zw(ackh, W. R. von H. 1826–1903)

Herbaria

ABD: Department of Botany, The University, Aberdeen, AB9 2UD, Scotland.
AL: Laboratoire de Botanique de la Faculté des Sciences, Université d'Alger, Alger, Algeria.
AMD: Hugo de Vries – Laboratorium, Hortus Botanicus, Plantage Midden-laan 2a, Amsterdam-4, Netherlands.
ATCC: American Type Culture Collection, 12301 Parklawn Drive, Rockville, Maryland 20852, U.S.A.
AWH: Natuurwetenschappelijk Museum van Antwerpen (Botanisch Museum Dr. H. van Heurck), Gerard Le Grellelaan 5, B 2020 Antwerpen, Belgium.
B: Botanisches Garten und Museum Berlin-Dahlem, Königin-Luise Strasse 6-8, D-1 Berlin 33 (Dahlem), Germany BRD. [For details of mycological collections not destroyed in World War II see Kohlmeyer (1962b) and for exsiccatae represented see Potztal (1961).]
BAS: Botanisches Institut der Universität Basel, Basel, Switzerland.
BERN: Botanisches Institut und Garten der Universität, Altenbergrain 21, CH-3013 Bern, Switzerland.
BG: Universitetets Botaniske Museum, Postboks 12, M-5014 Bergen, Norway.
BISH: Bernice P. Bishop Museum, Honolulu 17, Hawaii 96819, U.S.A.
BM: Cryptogamic Herbarium, Department of Botany, British Museum (Natural History), Cromwell Road, London SW7 5BD, England. [For information on the botanical collections see British Museum (Natural History) (1904). Most fungi formerly here are now in K.]
BO: Herbarium Bogoriense, Jalan Raya Juanda 22-24, Bogor, Indonesia.
BOL: Bolus Herbarium, The University, Rondebosch, Cape Town, South Africa.
BOLO: Istituto Botanico dell'Università, Via Irnerio 42, Bologna, Italy.
BONN: Botanisches Institut der Universität Bonn, Meckenheimer Allee 170, D-53 Bonn, Germany BRD.
BP: Department of Botany, Magyar Nemzeti Múzeum, Természettudo-mányi Múzeum, XIV, Várattiget, Budapest, Hungary. [For details of type specimens of lichens here see Verseghy (1964, 1968).]
BPI: The National Fungus Collections, Crops Research Division, United States Department of Agriculture, Beltsville, Maryland 20705, U.S.A. [For information on the collections see Lentz and Lentz (1968).]
BR: Jardin botanique national de Belgique, Domaine de Bouchout, B 1860 Meise, Belgium.
BREM: Übersee-Museum, Bahnhofsplatz 13, Brenem, Germany BRD.
BRIST: Department of Botany, The University, Bristol BS8 1UG, England.
BRU: Department of Botany, Brown University, Providence, Rhode Island, U.S.A.
BSB: Institut für Systematische Botanik und Pflanzengeographie der Freien Universität Berlin, Grunewaldstrasse 35, 1 Berlin 41 (Steglitz), Germany BRD.
BUCA: Institutul de Biologie "Tr. Săvulescu" al Academici R.P.R., Spl. Independentei 296, Bucuresti 17, Roumania.

Burnley: Plant Research Laboratory, Department of Agriculture Victoria, Burnley Gardens, Swan Street, Burnley E.1, Victoria, Australia.

C: Botanical Museum and Herbarium, Gothersgade 130, Copenhagen K, Denmark. [Some fungi have been transferred to the Institut for Sporenplanter, Copenhagen; see Lundqvist (1972) for notes on Hansen's material.]

CAN[L]: Lichen Herbarium, National Herbarium of Canada, National Museum of Natural Sciences, Ottawa, Ontario KIA 0M8, Canada.

CAS: California Academy of Sciences, Golden Gate Park, San Francisco, California 94118, U.S.A.

CBS: Centraalbureau voor Schimmelcultures, Oosterstraat 1, Baarn, Netherlands.

CM: Carnegie Museum, Carnegie Institute, 4400 Forbes Avenue, Pittsburgh, Pennsylvania 15213, U.S.A.

CN: Laboratoire de Botanique, Faculté des Sciences, Université de Caen, Caen, Calvados, France.

COI: Botanical Institute of the University of Coimbra, Coimbra, Portugal.

COLO: The Herbarium, University of Colorado Museum, Boulder, Colorado 80302, U.S.A.

Concarneau: Laboratoire de Biologie Marine du Collège de France, Concarneau, France. [Collections may not usually be borrowed.]

CP: Department of Plant Pathology, The Royal Veterinary and Agricultural College, Rolighedsvej 23, Copenhagen V, Denmark.

CRK: Department of Botany, University College, Cork, Eire. [For information on the lichen collections see Cullinane (1971).]

CUP: Department of Plant Pathology, Cornell University, Ithaca, New York 14850, U.S.A.

DAOM: National Mycological Herbarium, Biosystematics Research Institute, Canada Agriculture, Central Experimental Farm, Ottawa, Ontario K1A OC6, Canada.

DAR: New South Wales Department of Agriculture, Division of Science Services, P.O. Bag no. 10, Rydalmere, New South Wales, Australia.

DBN: The National Botanic Gardens, Glasnevin, Dublin, Ireland.

DS: Dudley Herbarium, Department of Biological Sciences, Stanford University, Stanford, California 94305, U.S.A.

DUKE: Department of Botany, Duke University, Durham, North Carolina 27706, U.S.A.

E: The Herbarium, Royal Botanic Garden, Inverleith Row, Edinburgh EH3 5LR, Scotland. [For information on the collections see Hedge and Lamond (1970).]

EA: The East African Herbarium, P.O. Box 5166, Nairobi, Kenya, East Africa. [Some lost?]

F: Field Museum of Natural History, Roosevelt Road at Lake Shore Drive, Chicago, Illinois 60605, U.S.A.

FH: Farlow Herbarium and Reference Library, Harvard University, 20 Divinity Avenue, Cambridge, Massachusetts 02138, U.S.A.

FI: Herbarium Universitatis Florentinae, Istituto Botanico, Via Lamarmora n. 4, 50121 Firenze, Italy.

FR: Forschungsinstitut und Naturmuseum Senckenberg, Senckenberg-Anlage 25, Frankfurt a.M., Germany BRD.

G: Conservatoire et Jardin Botanique, 192 route de Lausanne, Genève, Switzerland.

GA: Herbarium of the University of Georgia, Athens, Georgia, U.S.A.

GB: Botaniska Museet, Carl Skottsbergs gata 22, S-413 19 Göteborg, Sweden.

GE: Istituto ed Orto Botanico 'Hanbury' dell'Università, Corso Dogali I.C., Genova, Italy.

GENT: Laboratorium voor Plantensystematisk, 35 Ledeganckstraat, B-9000 Gent, Belgium.

GL: Department of Botany, The University, Glasgow W.2, Scotland.

GLAM: Department of Natural History, Glasgow Art Gallery and Museums, Kelvingrove, Glasgow, Scotland.

GOET: Systematisch-Geobotanisches Institut, Universität Göttingen, Unt. Karspüle 2, D 34 Göttingen, Germany BRD.

GRM: Muséum d'histoire Naturelle, 1 rue Dolomieu, Grenoble (Isère), France.

GRO: Afd. Plantensystematiek, Biologisch Centrum der Rijksuniversiteit te Groningen, Kerklaan 30, Postbus 14 Haren, Groningen, Netherlands. [Fungi formerly here now in L.]

H: Botanical Museum, University of Helsinki, Unioninkatu 44, SF 00170 Helsinki 17, Finland. [Specimens in the herbarium of Acharius may not be borrowed.]

HAL: Institut für Systematische Botanik und Pflanzengeographie der Martin-Luther-Universität, Neuwerk 21, Halle (Saale), Germany.

HBG: Staatsinstitut für Allgemeine Botanik und Botanischer Garten, Jungiusstrasse 6, Hamburg 36, Germany BRD.

HCIO: Herbarium cryptogamiae Indiae Orientalis, Division of Mycology and Plant Pathology, Indian Agricultural Research Institute, New Delhi 12, India.

HSI: Department of Silviculture, University of Helsinki, Unioninkatu 40B, Helsinki, Finland.

IA: Department of Botany, State University of Iowa, Iowa City, Iowa 52240, U.S.A.

IB: Institut für Botanik der Universität, Sternwartestrasse 15, Innsbruck, Austria.

ILL: Department of Botany, University of Illinois, Urbana, Illinois, U.S.A.

ILLS: Illinois State Natural History Survey, Natural Resources Building, Urbana, Illinois, U.S.A.

IMI: Commonwealth Mycological Institute, Ferry Lane, Kew, Richmond, Surrey TW9 3AF, England.

JE: Institut für Spezielle Botanik und Herbarium Haussknecht, Jena, Germany DDR.

K: The Herbarium, Royal Botanic Gardens, Kew, Richmond, Surrey TW9 3AE, England. [Most of the lichens formerly here are now in BM.]

Kaliningrad: Kaliningrad, Lithuanian SSR, U.S.S.R. [Full postal address unknown.]

KIEL: Botanisches Institut der Universität, Kiel, Germany BRD.

KRAM: Instytut Botaniki, Uniwersytet Jagielloński, ulica Lubicz 46, Kraków, Poland.

L: Rijksherbarium, Schelpenkade 6, Leiden, Netherlands.

LAM: Los Angeles County Museum, 900 Exposition Boulevard, Los Angeles, California 90007, U.S.A.

LAU: Musée botanique cantonal, Avenue de Cour 14 bis, CH 1007, Lausanne, Switzerland. .

LD: Universitets Botaniska Museum, Ö.Vallgatan 18, S-223 61 Lund, Sweden.

LE: Cryptogamic Herbarium, Komarov Botanical Institute of the Academy of Sciences of the U.S.S.R., Prof. Popov. Street 2, Leningrad P-22, U.S.S.R.

L'Harmas: Fabre Estate, L'Harmas, Sérignan, S. France. [See Lundqvist (1972, p. 285).]

LINN: Linnean Society of London, Burlington House, Piccadilly, London, W1V 0LQ. [For information on collections of Linnaeus see Howe (1911) and Vainio (1886a); and on other collections see Savage (1945). Specimens in the herbarium of Linnaeus may not be borrowed.]

LISU: Institute of Botany, Faculdade de Ciencias, Universidade de Lisboa, Lisboa 2, Portugal.

LPS: Instituto de Botánica C. Spegazzini, Calle 53 no. 477, La Plata, Bs. As., Argentina.

LUX: Musée d'Histoire Naturelle, Luxembourg, Luxembourg.

LZ: Botanisches Institut der Karl-Marx-Universität, Talstrasse 33, 701 Leipzig, Germany DDR. [Destroyed in World War II but new collections being built up.]

M: Botanische Staatssammlung, Menzingerstrasse 67, D-8 München 19, Germany BRD. [For details of Allescher's collections see Bresinsky (1973).]

MA: Instituto 'Antonio José Cavanilles', Jardin Botánico, Plaza de Murillo 2, Madrid 14, Spain.

MAF: Herbario del Laboratorio de Botánicade Facultad de Farmacia, Ciudad Universitaria, Madrid 3, Spain. [For details of Lázaro's collections see Wright and Calonge (1973).]

MAINE: Department of Botany, University of Maine, Orono, Maine, U.S.A.

MICH: University Herbarium, University of Michigan, Ann Arbor, Michigan 48104, U.S.A.

MIN: Department of Botany, University of Minneapolis, Minneapolis, Minnesota 55455, U.S.A.

MPU: Institut de Botanique Université de Montpellier, 5 rue Auguste Broussonnet, Montpellier (Hérault), France.

MSC: Department of Botany and Plant Pathology, Michigan State University, East Lansing, Michigan 48823, U.S.A.

MW: Department of Botany, Lomonosov State University, Leninskie Gory, Moscow, U.S.S.R. [For information on Hoffmann's collections see Vainio (1886b).]

NAP: Instituto Botanico della Università di Napoli, Via Foria 223, Napoli, Italy.

NCU: Herbarium of the University of North Carolina, Chapel Hill, North Carolina 27514, U.S.A.

NEB: University of Nebraska State Museum, Lincoln, Nebraska 68508, U.S.A.

NTM: Muséum d'histoire Naturelle de Nantes, Place de la Monnaie, Nantes, France.

NY: The New York Botanical Garden, Bronx Park, New York, New York 10458, U.S.A.

NYS: Herbarium, New York State Museum, Albany, New York 12201, U.S.A.

O: Botanisk Museum, Oslo 5, Norway.

ORE: Herbarium, University of Oregon, Eugene, Oregon, U.S.A.

OS: Department of Botany and Plant Pathology, Ohio State University, Columbus 10, Ohio, U.S.A.

OSC: Herbarium, Department of Botany and Plant Pathology, Oregon State University, Corvallis, Oregon 97331, U.S.A.

OXF: Department of Botany, The University, South Parks Road, Oxford OX1. [For information on the Dillenian collections see Crombie (1880) and for general information on the herbarium see Druce and Vines (1907).]

P: Laboratoire de Phanérogamie, 16 rue de Buffon, Paris 75005, France. [Some historical collections not split.]

PAC: Pennsylvania State University, University Park, Pennsylvania 16802, U.S.A.

PACA: Herbarium Anchieta, Instituto Anchietano e Unisinos, 93.000 São Leopoldo, Rio Grande do Sul, Brazil.

PAD: Istituto di Botanica e Fisiologia Vegetale, Università di Padova, Via Orto Botanico 15, 35100 Padova, Italy. [For details of Saccardo's herbarium see Gola (1930).]

PARMA: Istituto Botanico dell'Università, Parma, Italy.

PAV: Botanical Institute and Italian Cryptogamic Laboratory, The University, P.O. Box 99, 27100 Pavia, Italy.

PC: Laboratoire de Cryptogamie, Muséum National d'Histoire Naturelle, 12 rue de Buffon, Paris 75005, France. [For further information on the collections see Museum National d'Histoire Naturelle (1954).]

PDD: Auckland Plant Diseases Division, Department of Scientific and Industrial Research, Private Bag, Auckland, New Zealand.

PH: Academy of Natural Sciences, Philadelphia, Pennsylvania 19103, U.S.A.

PO: Instituto de Botânica 'Dr Gonçalo Sampaio', Faculdade de Ciencias, Universidade do Porto, Campo Alegre 1191, Porto, Portugal.

PR: Botanical Department, National Museum, Pruhonice, Praha, Czechoslovakia. [For details of Corda's herbarium see Pilát (1938).]

PRC: Universitatis Carolinae facultatis scientia naturalis cathedra, Benátska 2, Praha 2, Czechoslovakia.

PRE: Chief Plant Protection Research Institute, Agriculture Building, Beatrix Street, Private Bag X101, Pretoria, South Africa.

PUL: Kriebel Herbarium, Department of Biological Sciences, Purdue University, Lafayette, Indiana, U.S.A.

PUR: Arthur Herbarium, Department of Botany and Plant Pathology, Purdue University, Lafayette, Indiana 47907, U.S.A. [For further information see Baxter and Kern (1962).]

RO: Istituto Botanico, Città Universitaria, Roma, Italy.

RPPR: Institute of Tropical Forestry, P.O. Box AQ, Rio Piedras, Puerto Rico 00928.

S: Botanical Department, Naturhistoriska Riksmuseet, S-104 05 Stockholm, Sweden.

SAM: South African Museum Herbarium, National Botanic Gardens of South Africa, Kirstenbosch, Newlands, Cape Town, South Africa.

SIENA: Istituto Botanico della Università, Via P. A. Mattioli 4 (S. Agostino), 53100 Siena, Italy.

STE[-U]: Departement van Plantkunde, Universiteit van Stellenbosch, Stellenbosch, South Africa.

STR: Institut de Botanique de la Faculté des Sciences, 8 rue Goethe, 67-Strasbourg, France. [For further information on the collections see Kapp (1959).]

TAI: The Herbarium, Department of Botany, National Taiwan University, Taipei, Taiwan, China.

TENN: The Herbarium, University of Tennessee, Knoxville, Tennessee 37916, U.S.A. [Pre-1934 destroyed by fire.]

TL: Laboratoire de Botanique, Faculté des Sciences, Allées Jules Guesde, Toulouse (Haute Garonne), France.

TNS: Department of Botany, National Science Museum, Hyakunin-cho 3-23-1, Shinjuku-ku, Tokyo, Japan.

TO: Istituto Botanica della Università, Viale Mattioli 25, Torino 10125, Italy.

TOM: Istituto Missioni della Consolata, Corso Ferrucci 14, 10138 Torino, Italy.

TRH: Botanical Department, Museum of the Royal Norwegian Society for Science and Letters, 7000 Trondheim, Norway.

TRTC: Department of Botany, University of Toronto, Toronto, Ontario, Canada.

TUR: Botanical Institute of the University, Turku 2, Finland.

U: Botanical Museum and Herbarium, Tweede Transitorium, de Uithof, Utrecht, Netherlands.

UC: Herbarium of the University of California, Department of Botany, University of California, Berkeley California 94720, U.S.A.

UPS: Institute of Systematic Botany, The University, P.O. Box 541, S-751 21 Uppsala, Sweden.

UPSV: Växtbiologiska Institutionen, Uppsala Universitet, S-751 22 Uppsala, Sweden.

URM: Instituto de Micologia, Universidade do Recife, Avenida Rosa e Silva 347, 50.000 Recife, Pernambuco, Brazil.

US: Department of Botany, United States National Museum, Smithsonian Institution, Washington, D.C. 20560, U.S.A.

VER: Sezione di Botanica, Museo Civico di Storie Naturale, Corso Cavour II, 37100 Verona, Italy. [For further information on the collections see Bianchini (1968).]

W: Botanische Abteilung, Naturhistorisches Museum, Burgring 7, Postfach 417, A-1014 Wien, Austria.

WELT: The Dominion Museum, Wellington, New Zealand.
WIN: Herbarium of the Department of Botany, University of Manitoba, Winnipeg 1, Manitoba, Canada.
WIS: Herbarium of the University of Wisconsin, Madison, Wisconsin 53706, U.S.A.
WRSL: Instytut Botaniczny, Universytetu Wrocławskiego, ul. Kanonia 6/8, Wrocław, Poland.
WSP: Department of Plant Pathology, Washington State University, Pullman, Washington 99163, U.S.A.
WU: Botanisches Institut der Universität Wien, Rennweg 14, Wien III, Austria.
Z: Botanischer Garten und Institut für Systematische Botanik der Universität, 40 Pelikanstrasse, Zürich, Switzerland.
ZT: Institut für Spezielle Botanik, Eidgenossische Technischen Hochschule, Universitätsstrasse 2, CH-8006 Züric⁻, Switzerland.

VII. TITLE ABBREVIATIONS OF PUBLICATIONS

WHILE the titles of journals issued since 1900 cited in lists of synonyms may be abbreviated according to the fourth edition of the *World List of Scientific Periodicals* (see pp. 67, Fig. 6–7, 111, 114–115), there is no universally accepted system for abbreviating the titles of books or discontinued serials. Many works frequently cited in bibliographic citations of names have their titles abbreviated in a variety of ways. In an attempt to reduce this variation this section lists some of the more important works together with a few discontinued serials and floras and suggests suitable abbreviations.

For abbreviations of mycological books which are not included here see Commonwealth Mycological Institute (1969). It should be noted that where particular works have been issued over a number of years the dates given are for the whole period over which they appeared; the original works should be consulted to ascertain the dates of publication of particular parts or pages (see also pp. 110–111). Those which are discussed by Stafleu (1967) are followed by an asterisk (*).

The abbreviated title is always used in conjunction with the abbreviated author citation (see pp. 179–187) in the form 'Alb. & Schw., *Consp. Fung.*: 123 (1805)'. Volume, part and edition numbers should also be inserted where necessary.

Books

Acharius, E. (1798) *Lichenographiae svecicae Prodromus*. Linköping.   *Lich. suec. Prod.*

Acharius, E. (pre-April 1803) *Methodus qua omnes detectos Lichenes*. Stockholm.*   *Meth. Lich.*

Acharius, E. (1810) *Lichenographia universalis*. Göttingen.*   *Lich. univ.*

Acharius, E. (1814) *Synopsis methodica Lichenum*. Lund.   *Syn. Lich.*

Albertini, J. B. von and Schweinitz, L. D. von (1805) *Conspectus Fungorum*. Leipzig.   *Consp. Fung.*

Allescher, A. (1886–91) *Verzeichniss in Südbayern beobachteter Pilze*. Munich.   *Südbay. Pilze*

Batsch, A. J. G. C. (1783–89) *Elenchus Fungorum*. Halle and Salle.   *Elench. Fung.*

Battara, G. A. (1755) *Fungorum agri Ariminensis historia*. Faenza. [Re-issued in 1759.]*   *Fung. Arim.*

Berkeley, M. J. (1857) *Introduction to Cryptogamic Botany*. London.   *Intr. Crypt. Bot.*

Berkeley, M. J. (1860) *Outlines of British Fungology*. London.   *Outl. Br. Fung.*

Berlese, A. N. (1885–89) *Fungi moricolae*. Padua.   *Fungi moric.*

Berlese, A. N. (1894–1900) *Icones Fungorum omnium hucusque cognitorum*. Abellini and Padua.   *Icon. Fung.*

Bolton, J. (1788–91) *An History of Fungusses growing about Halifax*. 4 vols. Halifax and Huddersfield, etc.   *Hist. Fung.*

193

Bonorden, H. F. (1851) *Handbuch der allgemeinen Mykologie.*   *Handb. Mykol.*
Stuttgart.
Boudier, J. L. É. (1904–09) *Icones mycologicae.* Paris.   *Icon. mycol.*
Brefeld, O. (1872–1912) *Untersuchungen aus dem Gesammtgebiete*   *Unters. Mykol.*
der Mykologie. Leipzig. [Issued in 15 parts.]
Bresadola, G. (1881–92) *Fungi Tridentini novi vel nondum delineati*   *Fungi Trid.*
descripti et iconibus illustrati. Trento.
Bresadola, G. (1927–60) *Iconographia mycologica.* 28 vols. Milan.   *Icon. mycol.*
Brongniart, A. T. (1825) *Essai d'une Classification naturelle des*   *Essai Class.*
Champignons. Paris and Strasbourg.*   *Champ.*
Bulliard, J. B. F. [P.] (1780–98) *Herbier de la France.* 13 vols.   *Herb. Fr.*
Paris.*
Bulliard, J. B. F. [P.] (1791–1812) *Histoire des Champignons de*   *Hist. Champ. Fr.*
la France. 2 vols. Paris.*
Ceruti, A. (1948) *Fungi analytice delineati.* Turin.   *Fungi anal. del.*
Chevallier, F. F. (1826) *Flore générale des environs de Paris*, 1.   *Fl. env. Paris*
Paris.*
Chevallier, F. F. (1837) *Fungorum et Byssorum illustrationes.* Leip-   *Fung. Byss. ill.*
zig.
Clements, F. E. and Shear, C. L. (1931) *The Genera of Fungi.* New   *Gen. Fungi*
York.
Cooke, M. C. (1857–79) *Mycographia, seu Icones fungorum.* Lon-   *Mycograph.*
don. [Issued in parts.]
Cooke, M. C. (1871) *Handbook of British Fungi.* London and New   *Handb. Br. Fungi*
York.
Cooke, M. C. (1881–91) *Illustrations of British Fungi.* 8 vols. Lon-   *Ill. Br. Fungi*
don.
Corda, A. K. J. (1837–54) *Icones fungorum hucusque cognitorum.*   *Icon. fung.*
6 vols. Prague.*
Corda, A. K. J. (1839) *Pracht-flora europäischer Schimmelbil-*   *Pracht-fl.*
dungen. Leipzig and Dresden.*
Corda, A. K. J. (1842) *Anleitung zum Studium der Mycologie.*   *Anl. Stud. Mycol.*
Prague.
Cordier, F. S. (1870) *Les Champignons de la France.* Paris. [2 parts.]   *Champ. Fr.*
Crombie, J. M. (1894) *A Monograph of Lichens found in Britain*, 1.   *Monogr. Lich. Br.*
London. [See A. L. Smith for 2.]
Crouan, P. L. and Crouan, H. M. (1867) *Florule du Finistère.* Paris   *Fl. Finist.*
and Brest.
Dietrich, D. (1843–49) *Deutschlands kryptogamische Gewächse*   *Deut. krypt. Gew.*
(*Schwämme*). Jena.
Dillenius, J. J. (1741) *Historia muscorum.* Oxford.* [Also re-issued   *Hist. musc.*
(in part) in 1763, 1768 and 1811.]
Duby, J. E. (1828, 1830) *Aug. Pyrami de Candolle Botanicon galli-*   *Bot. gall.*
cum. 2 vols. Paris.*
Dumortier, B. C. J. (July–December 1822) *Commentationes botani-*   *Comm. bot.*
cae. Tournay. [Some copies dated 1823.]*
Ehrenberg, C. G. (1818) *Sylvae mycologicae Berolinenses.* Berlin.   *Sylv. mycol.*
Ehrhart, F. (1787–92) *Beiträge zur Naturkunde.* 7 vols. Hannover   *Beitr. Naturk.*
and Osnabrück.*
Ellis, J. B. and Everhart, B. M. (1892) *The North American*   *N. Am. Pyren.*
Pyrenomycetes. Newfield.
Eschweiler, F. G. (1824) *Systema Lichenum.* Nuremburg.   *Syst. Lich.*
Fée, A. L. A. (1824–25) *Essai sur les Cryptogames des écorces*   *Essai Crypt.*
exotiques officinales. Paris. [Issued in 7 parts.]*
Fée, A. L. A. (1837) *Essai sur les Cryptogames des écorces exotiques*   *Essai Crypt. Suppl.*
officinales, II. *Supplément et révision.* Paris.*
Ficinus, H. D. A. and Schubert, C. (1823) *Flora der Gegend um*   *Fl. Geg. Dresd.*
Dresden, II. *Kryptogamie.* Dresden.
Fresenius, J. B. G. W. (1850–63) *Beiträge zur Mykologie.* Frank-   *Beitr. Mykol.*
furt. [Issued in 3 parts.]*

Fries, E. M. (1815–1818) *Observationes mycologicae.* 2 vols.   *Obs. mycol.*
Copenhagen.
Fries, E. M. (1821–32) *Systema mycologicum.* 3 vols. Lund and   *Syst. mycol.*
Griefswald. [**1**, 1821; **2**(1), 1822; **2**(2), 1823; **3**(1), 1829;
3 (2, Index), 1832; see Art. 13 (f).]*
Fries, E. M. (1825) *Systema orbis vegetabilis.* Lund.   *Syst. orb.*
Fries, E. M. (1828) *Elenchus fungorum.* 2 vols. Greifswald. [see   *Elench. fung.*
Art. 13 (f).]*
Fries, E. M. (1831) *Lichenographia Europaea reformata.* Lund.   *Lich. Eur.*
Fries, E. M. (1836–38) *Epicrisis systematis mycologici.* Uppsala   *Epicr.*
and Lund.
Fries, E. M. (1846, 1849) *Summa vegetabilium Scandinaviae.* 2 vols.   *Summ. veg. Scand.*
Stockholm and Leipzig.
Fries, E. M. (1861–68) *Sveriges ätliga och giftiga Svampar.* Stock-   *Sver. atl.*
holm.
Fries, E. M. (1874) *Hymenomycetes Europaei.* Uppsala.   *Hymen. Eur.*
Fries, E. M. (1877) *Icones selectae Hymenomycetum.* 2 vols. Stock-   *Icon. Hymen.*
holm.
Fries, Th. M. (1871, 1874) *Lichenographia Scandinavica.* 2 vols.   *Lich. Scand.*
Uppsala.
Fuckel, K. W. G. L. (1870) *Symbolae mycologicae.* Wiesbaden.   *Symb. mycol.*
[Dated '1869' but issued in 1870.]*
Fuckel, K. W. G. L. (1871–75) *Symbolae mycologicae, Nachträge.*   *Symb. mycol.*
3 vols. Wiesbaden.*   *Nachtr.*
Gmelin, J. F. (1791–92) *Caroli à Linné Systema Naturae,* **2**. Leip-   *Car. Linn. Syst.*
zig.*   *nat.*
Gray, S. F. (November 1821) *A Natural Arrangement of British*   *Nat. Arr. Br. Pl.*
Plants. 2 vols. London.*
Greville, R. K. (1822–28) *Scottish cryptogamic Flora.* 6 vols. Edin-   *Scot. crypt. Fl.*
burgh.*
Greville, R. K. (1824) *Flora edinensis.* Edinburgh.*   *Fl. edin.*
Harmand, J. (1905–13) *Lichens de France.* 5 vols. Paris.   *Lich. Fr.*
Hepp, P. (1853–67) *Flechten Europas.* 16 vols. [32 fascicles.]   *Flecht. Eur.*
Zurich.
Hoffmann, G. F. (1784) *Enumeratio Lichenum.* Erlangen.   *Enum. Lich.*
Hoffman, G. F. (1789/90–1801) *Descriptio et adumbratio Plantarum*   *Descr. adumb.*
e classe cryptogamica Linnaei, quae Lichenes dicuntur. 3 vols.   *Lich.*
Leipzig. [Issued in parts.]
Hoffmann, G. F. (1796) *Deutschlands Flora oder botanisches*   *Deut. Fl.*
Taschenbuch, **2** *Kryptogamie.* Erlangen.*
Hoffmann, H. (1861–65) *Icones Analyticae Fungorum.* Giessen.   *Icon. fung.*
Hooker, W. J. (May 1821) *Flora scotica.* London. [Two parts.]*   *Fl. scot.*
Hudson, G. (1762) *Flora anglica.* London. [Ed. 2, 1778; Ed. 3,   *Fl. angl.*
1798.]*
Jatta, A. (1900) *Sylloge Lichenum italicorum.* Trani.   *Syll. Lich. ital.*
Kalchbrenner, C. (1873–77) *Icones selectae Hymenomycetum*   *Icon. Hymen.*
Hungariae. Budapest. [Issued in four parts.]
Karsten, P. A. (1871–79) *Mycologia Fennica.* Helsinki.   *Mycol. Fenn.*
Karsten, P. A. (1879) *Rysslands, Finlands och den Skandinaviska*   *Hattsvamp.*
halföns Hattsvampar. Helsinki.
Karsten, P. A. (1885) *Icones selectae Hymenomycetum Fenniae.*   *Icon. Hymen.*
Helsinki.
Kickx, J. (1867) *Flore cryptogamique des Flandres.* Paris.   *Fl. crypt. Fland.*
Körber, G. W. (1855) *Systema lichenum Germaniae.* Breslau.   *Syst. lich. Germ.*
Körber, G. W. (1859–65) *Parerga lichenologica.* Breslau. [Issued in   *Parerg. lich.*
parts.]
Kuhner, R. and Romagnesi, H. (1953) *Flore analytique des Cham-*   *Fl. Champ.*
pignons supérieurs. Paris.
Kuntze, F. C. E. (1891–98) *Revisio Generum plantarum.* 3 vols.   *Rev. Gen. Pl.*
Leipzig.*

Kunze, G. and Schmidt, J. K. (1817–23) *Mykologische Hefte*. 2 vols. Leipzig.　　*Mykol. Hefte*

Lamarck, J. B. A. P. M. de (1783–1817) *Encyclopédie méthodique Botanique*. 13 vols. Paris.* [Vols. 5–8 and Suppl. 1–5 by J. L. M. Poiret.]　　*Encycl. méth. Bot.*

Lamarck, J. B. A. P. M. de and De Candolle, A. P. (1805) *Flore française*. Ed. 3, **2**. Paris.*　　*Fl. fr.*

Lange, J. E. (1935–40) *Flora agaricina Danica*. 5 vols. Copenhagen.　　*Fl. agar. Dan.*

Leighton, W. A. (1871) *The Lichen Flora of Great Britain, Ireland, and the Channel Islands*. Shrewsbury. [Ed. 2, 1872; Ed. 3, 1879.]　　*Lich. Fl. Br.*

Léveillé, J. H. (1855) *Iconographie des Champignons de Paulet*. Paris.*　　*Icon. Champ.*

Lightfoot, J. (1778) *Flora scotica*. 2 vols. London.* [Dated '1777' but issued in 1778.]　　*Fl. scot.*

Lind, J. (1913) *Danish Fungi*. Copenhagen.　　*Dan. Fungi*

Link, [J.] H. F. (1829–33) *Handbuch zur Erkennung der nutzbarsten und am häufigsten vorkommenden Gewächse*. 3 vols. Berlin.*　　*Handb. Erk. Gew.*

Linnaeus, C. (1753) *Species Plantarum*. 2 vols. Stockholm. [See Art. 13 (d, h).]*　　*Sp. Pl.*

Lister, A. and Lister, G. (1925) *A Monograph of the Mycetozoa*. Ed. 3. London.　　*Monogr. Mycet.*

Lloyd, C. G. (1898–1925) *Mycological Notes*. Cincinnati.　　*Mycol. Notes*

Luyken, J. A. (1809) *Tentamen Historiae Lichenum*. Göttingen.　　*Tent. Hist. Lich.*

Martius, C. F. P. von (1817) *Flora cryptogamica Erlangensis*. Nuremberg.*　　*Fl. crypt. Erlang.*

Massalongo, A. B. (1852) *Richerche sull'autonomia dei Licheni crostosi*. Verona.　　*Ric. Lich. Crost.*

Massalongo, A. B. (1854) *Neagenea Lichenum*. Verona.　　*Neag. Lich.*

Massalongo, A. B. (1855) *Alcuni Generi di licheni*. Verona.　　*Gen. lich.*

Massalongo, A. B. (1856) *Miscellanea Lichenologia*. Verona.　　*Misc. Lich.*

Massee, G. E. (1892–95) *British Fungus Flora*. 4 vols. London.　　*Br. Fung. Fl.*

Matruchot, L. (1892) *Recherches sur les Développements de quelques Mucédinées*. Paris.　　*Rech. Dév. Mucéd.*

Mérat, F. V. (June 1821) *Nouvelle Flore des environs de Paris*. Ed. 2. Paris.*　　*Nouv. Fl. env. Paris*

Meyer, G. F. W. (1825) *Die Entwickelung, Metamorphose und Fortpflanzung der Flechten*. Göttingen.　　*Entw. Flecht.*

Michaux, A. (March 1803) *Flora boreali-Americana*. 2 vols. Paris and Strasbourg.*　　*Fl. bor.-Am.*

Micheli, P. A. (1729) *Nova plantarum genera juxta Tournefortii methodum disposita*. Florence.*　　*Nova. pl. gen.*

Migula, W. (1910–34) *Kryptogamen-Flora von Deutschland, Deutsch-Österreich und der Schweiz*, **3** (1*4). Berlin.　　*Krypt.-Fl.*

Montagne, J. F. C. (1856) *Sylloge Generum Specierumque cryptogamarum*. Paris.　　*Syll. Gen. Sp. crypt.*

Mudd, W. (1861) *A Manual of British Lichens*. Darlington.　　*Man. Br. Lich.*

Necker, N. J. de (1790) *Elementa Botanica*. Neuwied. [Also issued in 1791.]*　　*Elem. Bot.*

Nees von Esenbeck, C. G. D. (1816) *Das System der Pilze und Schwämme*. Würzburg. [Some copies are dated 1817.]*　　*Syst. Pilze Schw.*

Nees von Esenbeck, C. G. D. and Henry, A. C. F. (1837) *Das System der Pilze*. Bonn. [A second part was issued in 1858.]*　　*Syst. Pilze*

Nitschke, T. (1867) *Pyrenomycetes Germanici, Die Kernpilze Deutschlands*, **1**. Breslau.　　*Pyren. Germ.*

Nylander, W. (1858–60) *Synopsis methodica Lichenum*, **1**. Paris. [Vol. 2 was issued in 1863.]　　*Syn. Lich.*

Nylander, W. (1861) *Lichenes Scandinaviae*. Helsinki.　　*Lich. Scand.*

Nylander, W. (1890) *Lichenes Japoniae*. Paris.　　*Lich. Jap.*

Opiz, P. M. (1816) *Deutschlands cryptogamische Gewächse*. Prague. [Also issued in 1817 from Leipzig.]*　　*Deut. crypt. Gew.*

Oudemans, C. A. J. A. (1892–97) *Révisions des Champignons des Pays-Bas.* Amsterdam. — *Rev. Champ. Pays-Bas*

Oudemans, C. A. J. A. (1918–24) *Enumeratio systematica Fungorum.* 5 vols. The Hague — *Enum. syst. Fung.*

Patouillard, N. T. (1883–89) *Tabulae analyticae Fungorum.* Paris. — *Tab. anal. Fung.*

Patouillard, N. T. (1887) *Les Hyménomycètes d'Europe.* Paris. — *Hymén. Eur.*

Patouillard, N. T. (1900) *Essai taxonomique sur les familles et les genres des Hyménomycètes.* Lons-le-Saunier.* — *Essai Hymén.*

Penzig, O. (1887) *Funghi agrumicoli.* Padua. — *Fung. agrum.*

Persoon, C. H. (1796, 1799) *Observationes mycologicae.* 2 vols. Leipzig.* — *Obs. mycol.*

Persoon, C. H. (1797) *Tentamen dispositionis methodicae Fungorum.* Leipzig.* — *Tent. disp. meth. Fung.*

Persoon, C. H. (1798, 1800) *Icones et descriptiones Fungorum minus cognitorum.* 2 vols. Leipzig.* — *Icon. descr. Fung.*

Persoon, C. H. (1801) *Synopsis methodica Fungorum.* Göttingen. [See Art. 13 (e).]* — *Syn. meth. Fung.*

Persoon, C. H. (1803–08) *Icones pictae specierum rariorum Fungorum.* 4 vols. Paris.* — *Icon. pict. rar. Fung.*

Persoon, C. H. (1818) *Traité sur les Champignons comestibles.* Paris.* — *Traité Champ. comest.*

Persoon, C. H. (1822–28) *Mycologia europaea.* 3 parts. Erlangen.* — *Mycol. eur.*

Quélet, L. (1886) *Enchiridion fungorum in Europa Media et praesertim in Gallia vigentium.* Paris. — *Ench. fung.*

Quélet, L. (1888) *Flore mycologique de la France et des Pays limitrophes.* Paris. — *Fl. mycol. Fr.*

Rabenhorst, G. L. (1844–48) *Deutschlands Kryptogamenflora.* 2 vols. Leipzig. [Vol. 2 issued in parts.]* — *Deut. Kryptfl.*

Rea, C. (1922) *British Basidiomycetae.* Cambridge. — *Br. Basid.*

Rebentisch, J. F. (1804) *Prodromus Florae neomarchicae.* Berlin.* — *Prod. Fl. neom.*

Richon, C. and Roze, E. (1888) *Atlas des Champignons comestibles et vénéneux.* Paris. — *Atl. Champ.*

Saccardo, P. A. (1877–86) *Fungi italici.* 38 fascicles. Padua.* — *Fung. ital.*

Saccardo, P. A. (1882–1931, 1972) *Sylloge Fungorum.* 26 vols. Padua, etc. [With various collaborators.] — *Syll. Fung.*

St.-Amans, J. F. B. de (April 1821) *Flore agénaise.* Agen.* — *Fl. agén.*

Sartory, A. and Maire, L. (1922–23) *Compendium Hymenomycetum.* Paris. — *Compend. Hymen.*

Schaeffer, J. C. (1762–74) *Fungorum qui in Bavaria et Palatinatu circa Ratisbonam nascuntur.* 4 vols. Regensburg.* — *Fung. Bav.*

Schaerer, L. E. (1823–42) *Lichenum helveticorum spicilegium.* Bern. [Issued in parts.] — *Lich. helv. spic.*

Schaerer, L. E. (1850) *Enumeratio critica lichenum europaeorum.* Bern. — *Enum. lich. eur.*

Schlechtendal, D. F. L. von (1824) *Flora berolinensis*, 2 (*Cryptogamia*). Berlin.* — *Fl. berol. Crypt.*

Schreber, J. C. D. (1771) *Spicilegium florae lipsicae.* Leipzig.* — *Spic. fl. lips.*

Schreber, J. C. D. (1789, 1791) *Caroli a Linné Genera plantarum editio octava.* 2 vols. Frankfurt.* — *Car. Linn. Gen. pl.*

Schweinitz, L. D. von (September/October 1821) *Specimen florae Americae septentrionalis cryptogamicae.* Raleigh.* — *Sp. fl. Am. sept.*

Scopoli, G. A. (1772) *Flora carniolica.* Ed. 2, 2. Vienna. [Ed. 1 was issued in one volume in 1760.]* — *Fl. carn.*

Smith, A. L. (1911) *A Monograph of the British Lichens*, 2. London. [Vol. 1 by Crombie (q.v.); Ed. 2, 1 (1918), 2 (1926).] — *Monogr. Br. Lich.*

Smith, J. E. (1824–28) *The English Flora.* 4 vols. London. [Ed. 2, 5 vols. 1828–36; 5(1) 1833 by W. J. Hooker, 5(2) 1836 by M. J. Berkeley.]* — *Engl. Fl.*

Smith, J. E. and Sowerby, J. (1790–1814) *English Botany.* 36 vols. London. [Volumes issued in parts; all plates are dated.]* — *Engl. Bot.*

Sowerby J. (1796–1815) *Coloured figures of English Fungi or Mush-*   *Engl. Fungi*
rooms. 3 vols. and Suppl. London. [Issued in 32 parts.]*
Sprengel, K. P. J. (1806) *Florae halensis tentamen novum.* Halle.*   *Fl. halen.*
Sprengel, K. P. J. (1825–28) *Caroli Linnaei Systema vegetabilium*   *Car. Linn. Syst.*
   editio decima sexta. 5 vols. Göttingen.*   *veg.*
Steudel, E. G. (1824) *Nomenclator botanicus,* **2.** Stuttgart and   *Nomencl. bot.*
   Tübingen. [Ed. 2, 2 vols. (1840–41).]*
Stevenson, J. (1879) *Mycologia scotica.* Edinburgh.   *Mycol. scot.*
Stevenson, J. (1886) *Hymenomycetes Britannici.* 2 vols. Edinburgh   *Hymen. Br.*
   and London.
Sturm, J. (1813–62) *Deutschlands Flora,* 3 *Die Pilze Deutschlands.*   *Deut. Fl.*
   36 parts. Nuremburg. [Some parts by A. C. J. Corda, L. P. F.
   Ditmar, C. G. Preuss, etc.]*
Sydow, P. and Sydow, H. (1902–24) *Monographia Uredinearum.*   *Monogr. Ured.*
   4 vols. Leipzig.
Tode, H. J. (1790–91) *Fungi mecklenburgenses selecti.* Lüneburg.   *Fungi mecklenb.*
Tournefort, J. P. D. (1700) *Institutiones rei herbariae, editio altera.*   *Inst. rei herb.*
   Paris.*
Tuckerman, E. (1872) *Genera lichenum.* Amherst, Mass.   *Gen. lich.*
Tuckerman, E. (1882, 1888) *A Synopsis of the North American*   *Syn. N. Am. Lich.*
   Lichens. 2 vols. Boston and New Bedford, Mass.
Tulasne, L. R. (1851) *Fungi hypogaei.* Paris.   *Fungi hypog.*
Tulasne, L. R. and Tulasne, C. (1861–65) *Selecta Fungorum car-*   *Sel. Fung. carp.*
   pologia. 3 vols. Paris.
Villars, D. (1786–89) *Histoire des plantes de Dauphiné.* 3 vols.   *Hist. pl. Dauph.*
   Grenoble.*
Wallroth, K. F. W. (1831, 1833) *Flora cryptogamica Germaniae.*   *Fl. crypt. Germ.*
   2 vols. Nuremburg.*
Wiggers, F. H. (1780) *Primitiae florae holsaticae.* Kiel. [Mainly by   *Prim. fl. holsat.*
   G. H. Weber.]*
Willdenow, C. L. (1787) *Florae berolinensis Prodromus.* Berlin.*   *Fl. berol. Prod.*
Zahlbruckner, A. (1921–40) *Catalogus lichenum universalis.* 10 vols.   *Cat. lich. univ.*
   Leipzig. [Some volumes issued in parts.]

Discontinued journals

For details of other pre-1900 discontinued journals see British Museum (Natural History) (1968) and Lawrence *et al.* (1968).

Annalen der Botanick (ed. P. Usteri), Leipzig. 24 parts (1791–1800).   *Ulsteri's Annln*
   [Parts 7–24 entitled '*Neue Annalen*'.]*   *Bot.*
Grevillea, London. 22 vols. (1872–94).   *Grevillea*
Jahrbücher der Gewächskunde, Berlin and Leipzig. 1 vol. (1818–20).   *Jb. Gewächskde*
Journal of Botany (ed. W. J. Hooker), London. 4 vols. (1832–   *Hook. J. Bot.*
   1842).*
Linnaea (ed. D. F. L. von Schlechtendal), Berlin and Halle. 43 vols.   *Linnaea*
   (1826–82)*
Magazin für die Botanik (ed. J. J. Römer and P. Usteri), Zürich.   *Römer's Mag. Bot.*
   4 vols. (1787–91).*
Michelia (ed. P. A. Saccardo), Padua. 2 vols. (1877–82).   *Michelia*
Neues Journal für die Botanik (ed. H. A. Schrader), 4 vols. (1805–   *Schrad. Neues J.*
   1810).*   *Bot.*

Floras

Some floras issued in parts and containing contributions by different authors (and in some cases compiled by different editors) are most appropriately cited as if they were journals.

Flora italica cryptogama, Florence. [Fungi and lichens in 7 vols.   *Fl. ital. crypt.*
   (1908–34).]

Kryptogamen-Flora von Schlesien (ed. F. J. Cohn), Breslau. [Fungi and lichens constitute **2** (2)–**3** (1–2) (1879–1908); see *Taxon* **22**: 491–2 (1973).]   *Cohn's Krypt.-Fl. Schles.*

Kryptogamen-Flora der Mark Brandenburg, Leipzig. [Fungi and lichens constitute **5**–**9** (1905–57).]   *Krypt.-Fl. Brandenb.*

Dr L. Rabenhorst's Kryptogamen-Flora von Deutschland, Oesterreich und der Schweiz, Leipzig. [Fungi and lichens constitute **1** (1–8), **8**–**9** (1–5), 1885–1960; dates of publication for **1** (1–8) are on **1** (8): 852, 1907.]   *Rabenh. Krypt.-Fl.*

VIII. GLOSSARY

THIS section provides a glossary of some terms and abbreviations used in mycological nomenclature and is designed to assist in referring both to the Code and to
taxonomic publications. Some terms listed here are not now recommended or
have never been accepted in the Code and are included here only because they
have been employed in mycological publications. A detailed annotated bibliography of botanical nomenclature on which the present glossary has been partly
based has been published by McVaugh *et al.* (1968) and that work should be
consulted for terms used in the Code which are not, or only briefly, mentioned
here. Jeffrey (1973) includes a glossary of terms used in the nomenclature of
animals, bacteria, cultivated plants and viruses as well as in general botanical
nomenclature. A general dictionary of terms employed in microbial taxonomy
has been compiled by Cowan (1968) and those used in plant pathology are
discussed by Robinson (1969) and the Federation of British Plant Pathologists
(1973).

Terms denoting rank (and their abbreviations) have been discussed on pp.
38–47 and may be traced through the Index (pp. 223–231). Glossaries in which
the definitions of most terms used in the description of fungi and lichens may be
found have been referred to on p. 52.

absolute synonym: nomenclatural synonym.
accepted name: a name adopted by an author as the correct name for a taxon.
alternative names: two or more names in the same rank simultaneously published by an
    author for the same taxon, i.e. in the same work (see Art. 34).
analysis: an illustration showing the details necessary for the identification of a taxon
    (see Art. 41–42, 44).
apud: 'in'; sometimes used in citations when one author introduces a new name in the
    work of another (see Rec. 46D).
arithmotype: a specimen distributed as an isotype of a taxon (e.g. bearing the same
    collector's number) but belonging to a taxon which is different from the one to
    which the holotype belongs.
Article [Art.]: a rule of the Code which all botanists are required to follow.
asterisk (*): sometimes formerly used to indicate an infraspecific rank, frequently that of
    subspecies.
auctorum [auct.]: of authors; used in citing misapplied names.
auctorum anglicum [auct. angl.]: of English authors; see auctorum.
auctorum non [auct. non]: used in citing a name which has been misapplied (see Rec.
    50D).
author citation: the names of the author(s) responsible for introducing a name appended
    to the name itself (see Art. 46–49).
autonyms: automatically established names (see Art. 19, 22).
available name: a validly published legitimate name or epithet.
basinym: basionym.
basionym: the name bringing epithet in a new combination; i.e. the name on which a
    new combination is based (see Art. 33).
basonym: basionym.

binary name: a binomial name.

binomial: a name of two words composed of a generic name followed by a specific epithet (see Art. 23).

category: sometimes applied in the sense of taxonomic 'rank' or incorrectly used as synonymous with 'taxon' (see also Scott, 1973).

character: a feature of the organism under consideration.

circumscription: the diagnostic limits of a taxon.

cleptotype: a fragment from a holotype specimen.

Code: the most recent edition of the International Code of Botanical Nomenclature.

cohort, cohors: used by some nineteenth century authors as the term for a rank normally equivalent to that of 'order'.

combinatio nova [comb. nov.]: new combination; a new name that replaces an earlier one because of a change of rank or position which takes its epithet from the earlier name (basionym); used by authors making new combinations for the first time.

confamilial: belonging to the same family.

congeneric: belonging to the same genus.

conserved name: a validly published generic or family name whose retention has been authorized by an International Botanical Congress (see Art. 14) which would have to be replaced by a less well known name by the strict application of the Code; see rejected name.

conspecific: belonging to the same species.

correct name: a name that must be adopted for a particular taxon in a particular position under the Code (see Art. 6).

correctus [corr.]: corrected; used in citations of names to indicate who first corrected an orthographic error.

cotype: a term formerly employed for a 'syntype' but often misapplied to 'isotypes' and (or) 'paratypes'.

culto-type: a living culture derived from a holotype specimen.

cultype: a culture [incorrectly; see Art. 9] accorded the status of a nomenclatural type.

date of publication: in the Code the date of publication is the date on which a work was issued and may or may not be the same as the date printed on the work itself (see Art. 45).

description: an expanded diagnosis; often used as the equivalent of a diagnosis.

devalidated name: a name which would have been validly published except that it appeared before the starting-point date of the group to which its type belongs.

diagnosis: a statement of the characters of a newly described taxon which the author describing it consideres distinguishes it from other previously described taxa (see Art. 32).

discordant elements: elements on which a name was based which are now considered to belong to two or more distinct taxa (see Art. 70).

earlier (earliest, senior) homonym: the earliest validly published homonym of two or more homonyms.

earlier synonym: older synonym.

effective publication: publication in accordance with the Code (Art. 29–31).

emend. [emendatus, -a, -um]: used in author citations when an author has changed the circumscription of a taxon but has not excluded its nomenclatural type (see Rec. 47A).

ending: termination.

epithet: a word in a name of more than one word (other than a term denoting rank) following a generic name.

et [&]: and; used in the author citation of a name which two authors published jointly to link the names of the publishing authors, e.g. Berk. et Broome, Berk. & Broome (see Rec. 46B).

et al. [et alii, et aliorum]: and others; used in the author citation of a name to designate the second and other authors of a name when a name was published jointly by more than two authors, e.g. Fell *et al.* (see Rec. 46B).

etymology: the derivation and meaning of a name or epithet.

ex: from; used in the author citation of names to link the name of an author who has not validly published a name with that of the author who validated the same name, e.g. *Tomentella* Pers. ex Pat. (see Rec. 46C); sometimes used in a parallel way in herbarium abbreviations, e.g. BM ex K.

exclamation mark (!): sometimes used after a herbarium abbreviation (e.g. BM!) or collection to indicate that it has been examined or re-examined.

exclusis generibus/ speciebus/ varietatibus [excl. gen./sp./var.]: used after author citations of family/generic/specific names to indicate that the name is being used in a sense which does not include the genera/species/varieties placed in it by its describing author (see Rec. 47A).

exsiccata [plantae exsiccatae, fungi exsiccati]: sets of dried numbered herbarium specimens usually with printed labels distributed by sale, gift or exchange to botanical institutions.

facultative synonyms: taxonomic synonyms.

fascicle: a separately issued or bound part of a volume, exsiccatae, or serial publication (frequently called a 'number' in the case of journals).

fide: according to; teste.

generitype: the type species of a genus.

herbarium: a collection of preserved (usually dried) plant specimens, or the building in which such a collection is kept.

heterogeneous: used of a taxon which as circumscribed includes two or more elements which are considered to belong to one or more other taxa.

heterotypic synonyms: synonyms based on different nomenclatural types; taxonomic synonyms.

holotype [holotypus]: a single specimen or other single element used by an author or designated by him as the nomenclatural type of a taxon he is describing at the time of its original publication (see Art. 7); the 'nominifer' of zoologists.

homonym: a validly published name spelt exactly like another validly published name in the same rank but based on a different nomenclatural type (see Art. 64).

homotypic synonyms: nomenclatural synonyms.

illegitimate: used of a validly published name which is contrary to the Code and which must be rejected under it; see legitimate.

impriorable: illegitimate.

incertae sedis: of uncertain affinities; usually applied to family and generic names.

incidental mention: a name mentioned in a publication by an author who does not give any indication that he intends it to be adopted (see Art. 34).

indirect reference: a reference by the citation of an author's name or some other means to a previously published description of a taxon to which it refers other than by a complete or precise citation.

ineditatio [ined.]: unpublished.

infrageneric: used of any name in a rank below that of genus.

infraspecific: used of any name in a rank below that of species.

invalid: used of names which are not validly published according to the Code.

isoneotype: a duplicate of a neotype.

isonym: one of two or more names (i.e. new names or new combinations) based on the same nomenclatural type; see nomenclatural synonyms.

isoparatype: a duplicate of a paratype.

isosyntype: a duplicate of a syntype.

isotype ['isotypus']: a duplicate of a holotype, i.e. part of the single collection which includes the holotype (see Art. 7).

later synonyms: younger synonyms.

latinized: a name not of Latin origin but formed and treated according to the grammatical rules employed in that language.

lectotype [lectotypus]: a type selected from the original elements (specimens or names) on which a taxon was based when there is no holotype by a subsequent author to serve as the nomenclatural type of the name (see Art. 7).

legitimate: used of a validly published name which is in accordance with the Code (see Art. 6); see illegitimate.

loco citato [loc. cit.]: in the place cited; employed to avoid the repetition of a bibliographic reference.

locotype: topotype.

metonym: a later name given to a specimen (other than the type) of a taxon that has a valid name; i.e. a taxonomic synonym.

metonymous homonyms: homonyms based on different nomenclatural types which are considered to belong to a single taxon.

misapplied name: a name applied to a taxon that is different from the taxon to which the nomenclatural type of the name belongs.

misplaced term: a term denoting rank placed in an order contrary to that specified in Art. 3–5.

monodelphous homonyms: devalidated names taken up by different post-starting point authors independently but validated with different nomenclatural types.

monotype: a single element on which a taxon was based where the author did not explicitly state that this element was to be regarded as the nomenclatural type of the taxon; holotype.

monotypic: a genus or family which includes only a single species or genus, respectively.

monstrosity: a specimen showing an abnormal (non-hereditary) structural condition (see Art. 71).

mutatis characteribus [mut. char.]: emend.

name: as used in the Code unless otherwise indicated 'name' refers only to validly published names, whether they are legitimate or not (see Art. 12).

nec: not, nor; see non.

neotype: an element (usually a specimen) selected to serve as a nomenclatural type when all the original material on which a taxon was based has been lost or destroyed (see Art. 7).

new combination: see combinatio nova.

nomen abortivum [nom. abort.]: formerly applied to a name which at the date of publication was contrary to the Code in operation at that time.

nomen anamorphosis [nom. anam.]: a name based on an imperfect state of a pleomorphic fungus; stat. imperf.

nomen ambiguum [nom. ambig.]: a name used in different senses (i.e. applied to different taxa) that has become a persistent source of error.

nomen confusum [nom. conf.]: a name based on a nomenclatural type consisting of discordant elements.

nomen conservandum [nom. cons.] (see Rec. 50E): a conserved name (q.v.).

nomen conservandum propositum [nom. cons. prop.]: a family or generic name proposed for conservation but whose retention has not yet been authorized by an International Botanical Congress (see Art. 15); see nomen rejiciendum propositum.

nomen dubium [nom. dub.]: a name of uncertain application (e.g. because it cannot be satisfactorily typified).

nomen invalidum [nom. inval.]: a 'name' not validly published according to the Code; see name.

nomen monstrositatum [nom. monstr.]: a name based on a monstrosity (q.v.).

nomen neglectum: an overlooked but nevertheless validly published and legitimate name or epithet for a taxon; usually used of names not listed in catalogues of names (see pp. 108–109).

nomen non rite publicatum: a name not validly published according to the Code.

nomen novum [nom. nov.]: a new name introduced as a substitute for a previously validly published name based on the same nomenclatural type as the earlier name (see Art. 72).

nomen nudum [nom. nud.]: a 'name' published without a description or diagnosis and without any reference to a previously published description or diagnosis; consequently not a 'name' in the sense of the Code because it was not validly published (see Rec. 50B).

nomen provisorium [nom. provis.]: see provisional name.

nomen rejiciendum [nom. rejic.]: see rejected name; sometimes also applied to infrageneric names or epithets which have to be rejected under the Code because they are nomina ambigua, nomina confusa, nomina dubia, etc.

nomen rejiciendum propositum [nom. rejic. prop.]: a family or generic name proposed for rejection in favour of a name proposed for conservation but whose rejection has not yet been authorized by an International Botanical Congress; see nomen conservandum propositum.

nomen sed non planta [nom. non planta]: used in lists of synonyms to indicate that an author making a new combination applied the name to a taxon other than that to which the type of the basionym belongs.

nomenclatural status: the state of a name with a particular position and circumscription under the Code (i.e. validly published or not, legitimate or not, correct or not).

nomenclatural synonyms: two or more names based on the same nomenclatural type; sometimes indicated by '$\equiv$' in lists of synonyms (see Fig. 7B); see taxonomic synonyms.

nomenclatural type [typus]: the single element of a taxon to which its name is permanently attached, whether as a correct name or a nomenclatural synonym; i.e. the holotype, lectotype or neotype (as appropriate) of the name (Art. 7, 9).

nominifer: see holotype.

non: not; used alone in the citation of homonyms to indicate that they refer to taxa other than those described by the authors of the earlier homonyms (see Rec. 50C); see also 'auct. non', 'nec'.

objective synonyms: nomenclatural synonyms.

obligate synonyms: nomenclatural synonyms.

older synonym: an available earlier name in the same rank as the correct name for the taxon to which it belongs but which cannot be used for it under the Code.

opere citato [op. cit.]: in the work cited; loco citato.

ordo naturalis [ord. nat., nat. ord.]: natural order; used by some nineteenth century authors as the term for a rank equivalent to that of family (see Art. 18); see also 'cohors'.

original material: the elements (i.e. specimens, illustrations, descriptions, etc.) used by the author of a name of a taxon in the preparation of the protologue.

original spelling: the spelling of a name adopted by its author at the time of its valid publication, i.e. in the protologue (see Rec. 73E).

orthographic error: an unintentional spelling error; specified orthographic errors in names can be corrected by later authors without a change in the author citation (see Art. 73).

orthographic variants: alternative spelling forms of the same name (if these alternative spellings are based on different nomenclatural types they are treated as homonyms); also applied to names given to related taxa which are spelt so similarly that they are likely to be confused (see Art. 75).

orthography: spelling.

paranym: a validly published name so similar in orthography to another validly published name based on a different nomenclatural type that they are likely to be confused.

paratype: a specimen in addition to the holotype or isotype(s) mentioned in the original place of publication of a taxon (i.e. the protologue); also a remaining syntype after a lectotype has been chosen from the syntypes.

per: used in place of 'ex' in author citations by some authors where a pre-starting point name is being validated (see under Rec. 46E); see revalidated name.

pleomorphic: having many forms; used of a fungus having two or more distinct phases or states in its life cycle (see Art. 59).

pleonasm: the suggestion in a specific epithet of the meaning of a generic name.

polymorphic species: a species with many 'forms'; often used of species with many infraspecific taxa or morphotypes.

polynomial: a name in the rank of species, consisting of three or more separate words not hyphenated or separated by terms denoting rank, employed by most pre-Linnean authors (e.g. *Lichenoides pulmoneum reticulatum vulgare, marginibus peltiseris* Dill., *Hist. musc.*: 212, 1741) and in essence descriptive phrases; not 'names' in the sense of the Code as not in the binomial (binary) Linnean system; see binomial.

population: all individuals of one taxon growing in a particular region or place (usually considered as a single biological unit).

position: the place of a taxon as an element of the next higher taxon.

pre-Friesian: names or works published before 1 Jan. 1821 (see Art. 13).

pre-Linnaean: names or works published prior to 1 May 1753 (see Art. 13).

pre-starting point: names or works published prior to the starting point of the group to which the name or work belongs (see Art. 13).

priorable: legitimate.

priority: precedence by date of valid publication of a legitimate name over other names in the same rank applied to a particular taxon.

pro parte [p.p.]: in part; used in citations to indicate that the taxon in the sense of one author includes only a part of the taxon of the same name as understood by an earlier author.

proposal: a suggested addition or amendment to the Code to be considered by a subsequent International Botanical Congress.

pro synonymo [pro syn.]: used in citations to indicate that a name was originally published as a synonym of another (see Rec. 50A).

protologue: everything associated with a name when it was first published (see p. 136).

protonym: an effectively published name (but not devalidated or validly published) taken up and validly published at a later date.

provisional name: a name not accepted by an author when originally published but proposed in anticipation of the future acceptance of the taxon concerned (see Art. 34); see nomen provisorum.

publication: see date of publication, effective publication, validly published name.

rank: the position of a category in the taxonomic hierarchy.

recent: used in the Code in contradistinction to fossil.

recombination: sometimes applied to the placing of a name in an unaltered rank within the same genus or species but under a different taxon of higher rank (e.g. moving a 'form' from under one 'variety' of a species and placing it under another variety of the same species).

Recommendation: a part of the Code which is not obligatory but which botanists are urged to follow.

redeposition: the transference of a taxon to a new position with or without a change in the name or epithet.

rejected name: a family or generic name which by a strict application of the Code should replace a well known name but whose rejection in favour of the well known name has been authorized by an International Botanical Congress; see conserved name, nomen rejiciendum.

retain: to continue to use a name as the correct name or as the epithet of a correct name.

revalidated work: a devalidated name taken up and validly published in a post-starting point work; see per.

Rules: the criteria in the Code which must be followed set out in the Articles.

schizotype: a syntype which a later author implies he regards as the nomenclatural type of a taxon by excluding all other eligible syntypes from the taxon but does not specifically state he has selected a type for the taxon; i.e. a form of lectotype.

sensu [sens.]: in the sense of; used in author citations of misapplied names before the name of the author misapplying the name.

sensu amplo [sens. ampl.]: in the broad sense (see Rec. 47A).

sensu lato [sens. lat.]: in the wide sense; sensu amplo.

sensu stricto [sens. str.]: in the strict sense (see Rec. 47A).

sic: in this manner; used in citations in parentheses after a name or number to emphasize an orthographic or other error.

sine numero [s.n.]: sometimes used in citing lists of specimens after a collector's name to indicate the absence of a collector's number.

species nova [sp. nov.]: used after a name by an author describing a previously undescribed, or previously invalidly described, species for the first-time.

specific epithet: the second part of the binomial of a species name (the first being the name of the genus in which the species is placed).

square brackets [[ ]]: formerly recommended by the Code for including pre-starting point author citations (e.g. *Tomentella* [Pers.] Pat.); now replaced by 'ex' (e.g. *Tomentella* Pers. ex Pat.; see Rec. 46E); also sometimes used around a term denoting rank to indicate that the rank was not specified by the author of the name (e.g. *Alectoria jubata* [var.] *cana* Ach.).

standard: formerly used occasionally as equivalent to 'type'.

starting point: the date on which valid publication of names in a group begins (see Art. 13).

status: see nomenclatural status.

status conidialis [stat. conid.]: sometimes used in lists of synonyms to indicate that a name refers only to the conidial state of a pleomorphic fungus.

status imperfectus [stat. imperf.]: sometimes used in lists of synonyms to indicate that a name refers only to the imperfect state of a pleomorphic fungus.

status novus [*stat. nov.*]: used in taxonomic works to denote a name whose rank is being changed and where the epithet from the name in the old rank is being retained.

status pycnidialis [*stat. pycnid.*]: sometimes used in lists of synonyms to indicate that a name refers only to the pycnidial state of a pleomorphic fungus.

stem: the portion of a Greek or Latin word which is retained unchanged in the grammatical process of inflection.

subjective synonyms: taxonomic synonyms.

subordinate taxa: taxa considered to form a part or whole of a taxon of higher rank.

superfluous name: a name incorrectly applied to a taxon when another should have been adopted according to the Code (see Art. 63).

synisonym: one of two or more names having the same basionym; see nomenclatural synonym.

synonym [*synonymum*]: one of two or more names for the same taxon; see also 'nomenclatural synonym', 'taxonomic synonym'.

synonymum novum [*syn. nov.*]: sometimes used in lists of synonyms to indicate that a name is being treated as a taxonomic synonym of a particular accepted name for the first time.

synonymy: a list of names considered to be synonyms of the correct name of a taxon.

syntype: one of two or more elements cited by an author in the original place of publication of a taxon when no holotype was designated (see Art. 7).

tautonym: a specific or infraspecific name whose epithet repeats the generic name or correct name of the species (Art. 23, 27).

taxon vagum [*tax. vag.*]: a name used in a somewhat uncertain rank; 'tax. vag.' has sometimes been used in names in the position of and instead of a term denoting rank (e.g. *Usnea* [tax. vag.] *implexa* Hoffm.).

taxon [pl. *taxa*]: a taxonomic group of any rank; a 'taxon' is often used to include all its subordinate taxa (see Art. 25); see also 'category'.

taxonomic synonym: names based on different nomenclatural types but which a taxonomist considers refer to the same taxon; sometimes indicated by '=' in lists of synonyms (see Fig. 7B); see nomenclatural synonym.

term: a word used to designate a morphological or anatomical structure in contradistinction to 'name' as used in the Code.

termination: the part of a word added to a Greek or Latin stem when the word is inflected.

teste: according to; fide.

topotype: a specimen of a taxon from the same locality as the nomenclatural type of the taxon.

transfer: a change (with or without a change in name) in the position of a taxon.

type [*typ.*, *typus*]: see nomenclatural type.

type culture: a living culture derived from a holotype specimen; sometimes, incorrectly, a living culture on which a new taxon has been based.

type specimen: a type of a species or infraspecific taxon (see Art. 9).

typification: the process of designating or selecting a nomenclatural type for a name; see nomenclatural type.

typographic error: an error introduced by a printer (see Art. 73).

typonym: one of two or more names based on the same nomenclatural type as another name but which is neither its basionym nor a synisonym; nomenclatural synonyms.

typonymous homonyms: typonyms that are at the same time homonyms.

typotype: the type of the type; used where the type of a name is an illustration, the specimen from which the illustration was prepared being called the typotype.

typus conservandus [*typ. cons.*]: the nomenclatural type of a conserved generic name where the type is one other than that which would have to be adopted by a strict application of the Code but whose retention has been authorized by an International Botanical Congress.

uninomial: a unitary name.

unitary names: one word names applied to species by some eighteenth-century authors (see Art. 20).

unite: to treat as members of a single taxon elements which have previously been considered to belong to different taxa.

usage: the sense in which a name has been applied, whether correctly or incorrectly according to the Code.

validly published name: a name published according to the criteria for valid publication required by the Code (see Art. 32–44).

vernacular name: a colloquial name for a taxon in a non-classical language.

younger synonyms: available names in the same rank as the correct name of the taxon to which they belong but which are not themselves the correct name for it for reasons of priority.

IX. REFERENCES

Acharius, E. (1810) *Lichenographia universalis.* Göttingen.

Ahmadjian, V. (1967*a*) *The Lichen Symbiosis.* Waltham, Mass.: Blaisdell.

Ahmadjian, V. (1967*b*) A guide to the algae occurring as lichen symbionts: isolation, culture, cultural physiology, and identification. *Phycologia* 6: 127–160.

Ahmadjian, V. and Hale, M. E. (ed.) (1973) *The Lichens.* New York and London: Academic Press.

Ahmadjian, V. and Heikkilä, H. (1970) The culture and synthesis of *Endocarpon pusillum* and *Staurothele clopima. Lichenologist* 4: 259–267.

Ahti, T. (1961) Taxonomic studies on reindeer lichens (*Cladonia,* subgenus *Cladina*). *Ann. Bot. Soc. zool.-bot. fenn. Vanamo* 32(1): i-iv, 1–160.

Ainsworth, G. C. (1962) Pathogenicity and the taxonomy of fungi. *In* Ainsworth, G. C. and Sneath, P. H. A. (ed.) *Microbial Classification:* 249–269. Cambridge: Cambridge University Press.

Ainsworth, G. C. (1971) *Ainsworth & Bisby's Dictionary of the Fungi.* Ed. 6. Kew: Commonwealth Mycological Institute.

Ainsworth, G. C. (1973) Fungal nomenclature. *Rev. Pl. Path.* 52: 59–68.

Ainsworth, G. C., Sparrow, F. K. and Sussman, A. S. (ed.) (1973) *The Fungi, An advanced treatise* 4A–B. New York and London: Academic Press.

Ainsworth, G. C. and Sussman, A. S. (ed.) (1965, 1966, 1968) *The Fungi, An advanced treatise.* 3 vols. New York and London: Academic Press.

Alexopoulos, C. J. (1962) *Introductory Mycology.* Ed. 2. New York and London: Wiley.

Allorge, P. (1922) *Les associations végétales du Vexin français.* Dissertations, Nemours. [Not seen; cited by Barkman (1958).]

Ames, L. M. (1963) A monograph of the Chaetomiaceae. *U.S. Army Res. Dev. Ser.* 2: i-ix, 1–125.

Anon. (1971) Publish and be damned a second time. *Nature, Lond.* 233: 294.

Anon. (1972) International System of units. *N.Z. J. Bot.* 10: 507–512.

Apinis, A. E. (1972) Facts and problems. *Mycopath. Mycol. appl.* 48: 93–109.

Argus, G. W. and Sheard, J. W. (1972) Two simple labeling and data retrieval systems for herbaria. *Can. J. Bot.* 50: 2197–2209.

Arpin, N. (1969) Les caroténoïdes des Discomycètes: essai chimiotaxinomique. *Bull. Mens. Soc. Linn. Lyon* 38 Suppl.: 1–169.

Barkman, J. J. (1953) Comments on the rules of phytosociological nomenclature proposed by E. Meijer Drees. *Vegetatio* 4: 215–221.

Barkman, J. J. (1958) *Phytosociology and Ecology of Cryptogamic Epiphytes.* Assen: Van Gorcum.

Barkman, J. J. (1962) Bibliographia phytosociologica cryptogamica, Pars I: Epiphyta. *Excerpta Bot.* B, 4: 59–86.

Barkman, J. J. (1966) Bibliographia phytosociologica cryptogamica, Pars I: Epiphyta, Supplementum. *Excerpta Bot.* B, 7: 5–17.

Barkman, J. J. (1973) Taxonomy of cryptogams and cryptogam communities. *In* Heywood, V. H. (ed.) *Taxonomy and Ecology:* 141–150. London and New York: Academic Press.

Barnhart, J. H. (1965) *Biographical Notes upon Botanists.* 3 vols. Boston, Mass.: Hall.

Barrett, J. T. and Howe, M. L. (1968) Hemagglutination and hemolysis by lichen extracts. *Appl. Microbiol.* 16: 1137–1139.

Barron, G. L. (1971) Soil fungi. *In* Booth, C. (ed.) *Methods in Microbiology* 4: 405–427. London and New York: Academic Press.

Bartholomew, E. T. (1931) Herbarium arrangement of mycological specimens. *Mycologia* 23: 227–244.

Baxter, J. W. and Kern, F. D. (1962) History of the Arthur herbarium at Purdue University. *Proc. Indiana Acad. Sci.* **71**: 228–232.

Bellemère, A. (1967) Contribution à l'étude du développement de l'apothécie chez les Discomycètes Inoperculés. *Bull. Soc. mycol. Fr.* **83**: 393–640, 755–931.

Berkeley, M. J., Braithwaite, R., Cooke, M. C., Crombie, J. M. and Kitton, F. (1876) Priority of name. *Grevillea* **5**: 75.

Bertoldi, M. de, Lepidi, A. A. and Nuti, M. P. (1973) Significance of DNA base composition in classification of *Humicola* and related genera. *Trans. Br. mycol. Soc.* **60**: 77–85.

Bessey, E. A. (1950) *Morphology and Taxonomy of Fungi*. London: Constable.

Bianchini, F. (1968) Repertorio degli erbari conservati nella Sezione di Botanica del Museo Civico di Storia Naturale di Verona. *Memorie Mus. civ. Stor. nat. Verona* **15**: 447–449.

Bisby, G. R. (1953) *An Introduction to the Taxonomy and Nomenclature of Fungi*. Ed. 2. Kew: Commonwealth Mycological Institute.

Blake, S. F. and Atwood, A. C. (1942, 1961) *A Geographical Guide to Floras of the World*. 2 vols. Washington: U.S. Department of Agriculture [Publ. no. 401, 797.]

Blunt, W. (1950) *The Art of Botanical Illustration*. London: Collins.

Bohus, G. (1963) New suggestions for preparing fleshy fungi for the herbarium. *Mycologia* **55**: 128–130.

Bonner, C. E. B. (1962→) *Index Hepaticarum* I→. Weinheim: Cramer.

Booth, C. (ed.) (1971*a*) *Methods in Microbiology* 4. London and New York: Academic Press.

Booth, C. (1971*b*) Introduction to general methods. *In* Booth, C. (ed.) *Methods in Microbiology* **4**: 1–47. London and New York: Academic Press.

Booth, C. (1971*c*) Fungal culture media. *In* Booth, C. (ed.) *Methods in Microbiology* **4**: 49–94. London and New York: Academic Press.

Booth, C. (1971*d*) *The Genus* Fusarium. Kew: Commonwealth Mycological Institute.

Bottle, R. T. and Wyatt, H. V. (ed.) (1966) *The use of Biological Literature*. London: Butterworths.

Bourne, C. P. (1962) The world's technical journal literature: an estimate of volume, origin, language, field, indexing, and abstracting. *Am. Docum.* **13**: 159–168.

Bracker, C. E. (1967) Ultrastructure of fungi. *A. Rev. Phytopath.* **5**: 343–374.

Brenan, J. P. M. (1972) *Draft Index of Author Abbreviations: Flowering Plants*. Kew: Royal Botanic Gardens. [Mimeographed; a supplement was also prepared in 1973.]

Bresinsky, A. (1973) Typen-Liste der von Andreas Alleschen neu beschribenen Pilzsippen. *Mitt. bot. StSamml., Münch.* **11**: 33–55.

British Museum (Natural History) (1904) *The history of the Collections contained in the Natural History Department of the British Museum* **1**. London: British Museum (Natural History).

British Museum (Natural History) (1968) *List of Serial Publications in the British Museum (Natural History) Library*. London: British Museum (Natural History).

British Standards Institution (1958) *B.S. No. 1219, Recommendations for Proof Correction and Copy Preparation*. London: British Standards Institution.

British Standards Institution (1969) *The Use of SI Units*. London: British Standards Institution.

Brodo, I. M. (1968) The lichens of Long Island, New York: A vegetational and floristic analysis. *Bull. N.Y. St. Mus. Sci. Serv.* **410**: i–x, 1–330.

Brookes, B. C. (1971) Optimum P% library of scientific periodicals. *Nature, Lond.* **232**: 458–461.

Buchanan, R. E., Holt, J. G. and Lessel, E. F. (ed.) (1966) *Index Bergeyana*. Edinburgh and London: Livingstone.

Burges, A. (1958) *Micro-organisms in the Soil*. London: Hutchinson.

Burnett, J. H. (1968) *Fundamentals of Mycology*. London: Arnold.

Burton, A. L. (ed.) (1971) *Cinematographic Techniques in Biology and Medicine*. London and New York: Academic Press.

Cadbury, D. A., Hawkes, J. G. and Readett, R. C. (1971) *A Computer Mapped Flora*. London and New York: Academic Press.

Campbell, I. (1971) Numerical taxonomy of various genera of yeasts. *J. gen. Microbiol.* **67**: 223–231.

Campbell, I. S. C. (1968) Lichen photography. *Bull. Br. Lichen Soc.* **1**(23): 7–10.
Campbell, R. (1972) Ultrastructure of conidium ontogeny in the Deuteromycete fungus *Stachybotrys atra* Corda. *New Phytol.* **71**: 1143–1149.
Carlile, M. J. (1971) Myxomycetes and other slime moulds. *In* Booth, C. (ed.) *Methods in Microbiology* **4**: 237–265. London and New York: Academic Press.
Carmichael, J. W. and Sneath, P. H. A. (1969) Taxometric maps. *Syst. Zool.* **18**: 402–415.
Casey, R. S., Perry, J. W., Berry, M. M. and Kent, A. (ed.) (1958) *Punched Cards, their Applications to Science and Industry*. Ed. 2. New York: Reinhold.
Cash, E. K. (1965) *A Mycological English-Latin Glossary*. New York and London: Hafner.
Chadefaud, M. and Avellanas, L. (1967) Remarques sur l'ontogénie et la structure des périthèces des "*Chaetomium*". *Botaniste* **50**: 59–87.
Chaplin, A. H. (1967) *Names of Persons: National Usages for entry in Catalogues*. Definitive Ed. [Chaplin, A. H. and Anderson, D. (ed.).] Sevenoaks: International Federation of Library Associations.
Chaudri, M. N., Vegter, I. H. and de Wal, C. M. (1972) Index Herbariorum Part II(3), Collectors I–L. *Regnum Vegetabile* **86**: i-xxii, 297–473.
Chevalier, A. (1947) Les destructions causées par la guerre dans les grandes collections botaniques. *Revue internat. Bot. appl.* **27**: 37–53.
Christensen, C. (1906–65) *Index Filicum*. 5 vols. Copenhagen: Hagerup.
Ciferri, R. (1952) Localization of mycological type specimens in Italy. *Taxon* **1**: 126–127.
Ciferri, R. (1957) Type nomenclature of micro-organisms in culture. *Taxon* **6**: 154.
Ciferri, R. (1957–60) *Thesaurus literaturae mycologicae et lichenologicae, Supplementum 1911–1930*. 4 vols. Cortina: Papia.
Clements, F. E. and Shear, C. L. (1931) *The Genera of Fungi*. New York: Wilson.
Cole, A. J. (ed.) (1969) *Numerical Taxonomy*. London and New York: Academic Press.
Cole, G. T. and Kendrick, W. B. (1968) A thin culture chamber for time-lapse photomicrography of fungi at high magnifications. *Mycologia* **60**: 340–344.
Collins, F. H. (1956) *Authors' and Printers' Dictionary* Ed. 10. London: Oxford University Press.
Committee on Form and Style of the Council of Biology Editors (1972) *CBE Style Manual*. Ed. 3. Washington: American Institute of Biological Sciences.
Committee of the Library Association and Committee of the American Library Association (1965) *Cataloguing Rules: Author and Title Entries*. English Ed. London: Library Association.
Commonwealth Mycological Institute (1968) *Plant Pathologist's Pocketbook*. Kew: Commonwealth Mycological Institute.
Commonwealth Mycological Institute (1969) Title abbreviations for some common mycological taxonomic publications. *Bibl. Syst. mycol.* **4**, *Suppl.*: 1–13.
Commonwealth Mycological Institute (1971) *International Mycological Directory*. Kew: Commonwealth Mycological Institute.
Cooke, W. B. (1948) A survey of literature on fungus sociology and ecology. *Ecology* **29**: 376–382.
Cooke, W. B. and Hawksworth, D. L. (1970) A preliminary list of the families proposed for fungi (including the lichens). *Mycol. Pap.* **121**: 1–86.
Copeland, E. B. (1947) *Genera Filicum*. Waltham, Mass.: Chronica Botanica.
Copeland, H. F. (1956) *The Classification of Lower Organisms*. Palo Alto, Calif.: Pacific Books.
Cowen, R. S. (1970) The Index Nominum Genericorum project—past, present, and future. *Taxon* **19**: 52–54.
Cowen, S. T. (1968) *A Dictionary of Microbial Taxonomic Usage*. Edinburgh: Oliver and Boyd.
Crombie, J. M. (1880) On the lichens in Dillenius's 'Historia Muscorum' as illustrated by his herbaria. *J. Linn. Soc., Bot.* **17**: 553–581.
Culberson, C. F. (1969) *Chemical and Botanical Guide to Lichen Products*. Chapel Hill, N.C.: University of North Carolina Press.
Culberson, C. F. (1970) Supplement to 'Chemical and Botanical Guide to Lichen Products'. *Bryologist* **73**: 177–377.
Culberson, C. F. (1972) Improved conditions and new data for the identification of

lichen products by a standardized thin-layer chromatographic method. *J. Chromatog.* **72**: 113–125.

Culberson, C. F. and Kristinsson, H. (1970) A standardized method for the identification of lichen products. *J. Chromatog.* **46**: 85–93.

Culberson, W. L. (1951 →) Recent literature on lichens 1 →. *Bryologist* **54** →.

Culberson, W. L. (1969) The use of chemistry in the systematics of lichens. *Taxon* **18**: 152–166.

Culberson, W. L. and Culberson, C. F. (1970). A phylogenetic view of chemical evolution in the lichens. *Bryologist* **73**: 1–31.

Culberson, W. L. and Culberson, C. F. (1973) Parallel evolution in lichen-forming fungi. *Science, N. Y.* **180**: 196–198.

Cullinane, J. P. (1971) The lichen herbarium at University College, Cork. *Ir. Nat. J.* **17**: 45–49.

Dade, H. A. (1949) Colour terminology in biology. Ed. 2. *Mycol. Pap.* **6**: 1–22.

Dade, H. A. and Gunnell, J. (1969) *Classwork with Fungi.* Revised Ed. Kew: Commonwealth Mycological Institute.

Dahl, E. and Krog, H. (1973) *Macrolichens of Denmark, Finland, Norway and Sweden.* Oslo, etc.: Universitetsforlaget.

Davies, R. R. (1971) Air sampling for fungi, pollens and bacteria. *In* Booth, C. (ed.) *Methods in Microbiology* **4**: 367–404. London and New York: Academic Press.

Davis, P. H. and Heywood, V. H. (1963) *Principles of Angiosperm Taxonomy.* Edinburgh and London: Oliver and Boyd.

Dawson, E. Y. (1962) *New taxa of Benthic Green, Brown and Red Algae published since De Toni.* Santa Yuez, Calif.: Beaudette Foundation.

De Candolle, A. [P.] (1880) *La Phytographie.* Paris: Masson.

Degelius, G. (1954) The lichen genus *Collema* in Europe. *Symb. bot. upsal.* **13**(2): 1–499.

Deighton, F. C. (1960) Collecting fungi in the tropics. *In* Commonwealth Mycological Institute, *Herb. I.M.I. Handbook*: 78–83. Kew: Commonwealth Mycological Institute.

Dennis, R. W. G. (1968) *British Ascomycetes.* Lehre: Cramer.

Dennis, R. W. G. (1973) The fungi of southeast England. *Kew Bull.* **28**: 133–139.

De Toni, J. B. (1889–1924) *Sylloge Algarum.* 6 vols. Pavia.

Dibben, M. J. (1971) Whole-lichen culture in a phytotron. *Lichenologist* **5**: 1–10.

Dillenius, J. J. (1741) *Historia muscorum.* Oxford.

Dring, D. M. (1971) Techniques for microscopic preparation. *In* Booth, C. (ed.) *Methods in Microbiology* **4**: 95–111. London and New York: Academic Press.

Druce, C. G. and Vines, S. H. (1907) *The Dillenian Herbaria.* Oxford: Clarendon Press.

Duncan, E. G. and Galbraith, M. H. (1973) Improved procedures in fungal cytology using Giesma. *Stain Technol.* **48**: 107–110.

Duncan, U. K. (1970) *Introduction to British Lichens.* Arbroath: Buncle.

Durbin, R. D. (1966) Comparative gel-electrophoretic investigation of the protein patterns in *Septoria* species. *Nature, Lond.* **210**: 1186–1187.

Du Rietz, G. E. (1930) The fundamental units of biological taxonomy. *Svensk Bot. Tidskr.* **24**: 333–428.

Edwards, J. G. (1954) A new approach to infraspecific categories. *Syst. Zool.* **3**: 1–20.

Edwards, P. I. (1971a) List of abstracting and indexing services in pure and applied biology. *Biol. J. Linn. Soc.* **3**: 277–286.

Edwards, P. I. (1971b) List of libraries in the field of pure and applied biology. *Biol. J. Linn. Soc.* **3**: 173–188.

Eihellinger, A. (1973) Die Pilze der Pflanzengesellschaften des Auwaldgebiets der Isar zwischen München und Grüneck. *Ber. Bayer. bot. Ges.* **44**: 5–100.

Ellis, M. B. (1950) A modification of the 'Necol' technique for mounting microfungi. *Trans. Br. mycol. Soc.* **33**: 22.

Ellis, M. B. (1960) The herbarium. *In* Commonwealth Mycological Institute, *Herb. I.M.I. Handbook*: 24–36. Kew: Commonwealth Mycological Institute.

Ellis, M. B. (1971) *Dematiaceous Hyphomycetes.* Kew: Commonwealth Mycological Institute.

Engel, C. E. (ed.) (1968) *Photography for the Scientist.* London and New York: Academic Press.

Eriksson, B. (1970) On ascomycetes on Diapensiales and Ericales in Fennoscandia 1. Discomycetes. *Symb. bot. upsal.* **19**(4): 1–71.

Esdaile, A. (1957) *National Libraries of the World.* Ed. 2. London: Library Association.

Esser, K. and Keunen, R. (1965) *Genetik der Pilze.* Berlin, etc.: Springer.

Fairbrothers, D. E. (1968) Chemosystematics with emphasis on systematic serology. *In* Heywood, V. H. (ed.) *Modern Methods in Plant Taxonomy:* 141–174. London and New York: Academic Press.

Federation of British Plant Pathologists (1973) A guide to the use of terms in plant pathology. *Phytopath. Pap.* **17**: i–iv, 1–54.

Ferry, B. W., Baddeley, M. S. and Hawksworth, D. L. (ed.) *Air Pollution and Lichens.* London: University of London Athlone Press; Toronto: University of Toronto Press.

Fincham, J. R. S. and Day, P. R. (1971) *Fungal Genetics.* Ed. 3. Oxford and Edinburgh: Blackwell.

Fletcher, A. (1973) The ecology of marine (littoral) lichens on some rocky coasts of Anglesey. *Lichenologist* **5**: 368–400.

Forest Products Research Laboratory (1960) *Identification of Hardwoods, A lens key.* Ed. 2. [*F.P.R. Bull. no.* 25]. London: H.M.S.O.

Fosberg, F. R. and Sachet, M.-H. (1965) Manual for tropical herbaria. *Regnum Vegetabile* **39**: 1–132.

Fowler, H. W. (1965) *A Dictionary of modern English usage.* Ed. 2. Oxford: Clarendon Press.

Frey, E. (1959) Die Flechtenflora und -vegetation des Nationalparks im Unterengadin II. Die Entwicklung der Flechtenvegetation auf photogrammetrisch kontrollierten Dauerflächen. *Ergebn. wiss. Unters. schweiz. NatnParks* **6**: 241–319.

Fries, E. M. (1821) *Systema mycologicum* 1. Greifswald.

Galun, M., Marton, K. and Behr, L. (1972) A method for the culture of lichen thalli under controlled conditions. *Arch. Mikrobiol.* **83**: 189–192.

Gams, H. (1927) Von den Follatères zur Dent de Morcles. *Beitr. Geobot. Landesauf.* **15**: i–xii, 1–760.

Gates, B. N. (1958) A new soil-binder for preserving lichen specimens. *Bryologist* **61**: 249–252.

Gilbert, O. L. (1971) Studies along the edge of a lichen desert. *Lichenologist* **5**: 11–17.

Gilmour, J. (1973) Octal notation for designating physiologic races of plant pathogens. *Nature, Lond.* **242**: 620.

Gilmour, J. S. L. *et al.* (ed.) (1969) International code of nomenclature of cultivated plants—1969. *Regnum Vegetabile* **64**: 1–32.

Gola, G. (1930) *L'Herbario Mycologico di P. A. Saccardo, Catalogo.* Padova.

Goodall, D. W. (1952) Quantitative aspects of plant distribution. *Biol. Rev.* **27**: 194–245.

Goodall, D. W. (1953) Objective methods for the classification of vegetation II. Fidelity and indicator values. *Austr. J. Bot.* **1**: 434–456.

Gould, S. W. and Noyce, D. C. (1965) Authors of plant genera. *International Plant Index* **2**: 1–336.

Graesse, [J. G. T.], Benedict, [?] and Plechl, [H.] (1971) *Orbis latinus.* Ed. 4. Braunschweig: Klinkhardt and Biermann.

Greenhalgh, G. N. and Evans, L. V. (1971) Electron microscopy. *In* Booth, C. (ed.) *Methods in Microbiology* **4**: 517–565. London and New York: Academic Press.

Grieg-Smith, P. (1964) *Quantitative Plant Ecology.* Ed. 2. London: Butterworths.

Griffiths, H. B. (1973) Fine structure of seven unitunicate pyrenomycete asci. *Trans. Br. mycol. Soc.* **60**: 261–271.

Grove, W. B. (1937) *British Stem- and Leaf-Fungi (Coelomycetes)* **2**. Cambridge: Cambridge University Press.

Grummann, V. (1941) Morphologische, anatomische und entwicklungsgeschichtliche Studien über Bildungsabweichungen bei Flechten. *Reprium nov. Spec. Rengi veg., Beih.* **122**: 1–128.

Grummann, V. (1963) *Catalogus Lichenum Germaniae.* Ztuttgart: Gustav Fischer.

Grummann, V.† (1974) *Biographisch-bibliographisches Handbuch der Lichenologie.* Lehre: Cramer. [Not seen.]

Hale, M. E. (1967) *The Biology of Lichens.* London: Arnold.

Hale, M. E. (1969) *How to Know the Lichens.* Dubuque, Iowa: Brown.

Hale, M. E. (1972) *Parmelia pustulifera*, a new lichen from south-eastern United States. *Brittonia* **24**: 22–27.

Hale, M. E. (1973) Fine structure of the cortex in the lichen family Parmeliaceae viewed with the scanning electron microscope. *Smithsonian Contr. Bot.* **10**: 1–92.

Hall, A. V. (1970) A computer-based system for forming identification keys. *Taxon* **19**: 12–18.

Hall, A. V. (1972*a*) Computer-based data banking for taxonomic collections. *Taxon* **21**: 13–15.

Hall, A. V. (1972*b*) The use of a data-banking system for taxonomic collections. *Contr. Bolus Herb.* **5**: 1–78.

Hall, R. (1969) Molecular approaches to taxonomy of fungi. *Bot. Rev.* **35**: 285–304.

Harris, R. C. (1973) The corticolous pyrenolichens of the Great Lakes region. *Mich. Bot.* **12**: 3–68.

Hart, H. (1952) *Rules for Compositors and Readers at the University Press, Oxford.* Ed. 36. London: Oxford University Press.

Harter, L. L. (1941) The personal element and light as factors in the study of the genus *Fusarium. J. agric. Res.* **62**: 97–107.

Hawker, L. E. (1955) Hypogeous fungi. *Biol. Rev.* **30**: 127–158.

Hawksworth, D. L. (1969*a*) The lichen flora of Derbyshire. *Lichenologist* **4**: 105–193.

Hawksworth, D. L. (1969*b*) The scanning electron microscope. An aid to the study of cortical hyphal orientation in the lichen genera *Alectoria* and *Cornicularia. J. Microscopie* **8**: 753–760.

Hawksworth, D. L. (1970*a*) Guide to the literature for the identification of British lichens. *Bull. Br. mycol. Soc.* **4**: 73–95.

Hawksworth, D. L. (1970*b*) The chemical constituents of *Haematomma ventosum* (L.) Massal. in the British Isles. *Lichenologist* **4**: 248–255.

Hawksworth, D. L. (1971*a*) A brief guide to microchemical techniques for the identification of lichen products. *Bull. Br. Lichen. Soc.* **2**(28): 5–9.

Hawksworth, D. L. (1971*b*) A revision of the genus *Ascotricha* Berk. *Mycol. Pap.* **126**: 1–28.

Hawksworth, D. L. (1972*a*) Regional studies in *Alectoria* (Lichenes) II. The British species. *Lichenologist* **5**: 181–261.

Hawksworth, D. L. (1972*b*) The natural history of Slapton Ley Nature Reserve IV. Lichens. *Fld Stud.* **3**: 535–578.

Hawksworth, D. L. (1973*a*) Some advances in the study of lichens since the time of E. M. Holmes. *Bot. J. Linn. Soc.* **67**: 3–31.

Hawksworth, D. L. (1973*b*) Ecological factors and species delimitation in the lichens. *In* Heywood, V. H. (ed.) *Taxonomy and Ecology*: 31–69. London and New York: Academic Press.

Hawksworth, D. L. (1973*c*) Mapping studies. *In* Ferry, B. W., Baddeley, M. S. and Hawksworth, D. L. (ed.) *Air Pollution and Lichens*: 38–76. London: University of London Athlone Press; Toronto: University of Toronto Press.

Hawksworth, D. L. and Chapman, D. S. (1971) *Pseudevernia furfuracea* (L.) Zopf and its chemical races in the British Isles. *Lichenologist* **5**: 51–58.

Hawksworth, D. L. and Punithalingam, E. (1973) Typification and nomenclature of *Dichaena* Fr., *Heterographa* Fée, *Polymorphum* Chev., *Psilospora* Rabenh. and *Psilosporina* Died. *Trans. Br. mycol. Soc.* **60**: 501–509.

Hawksworth, D. L. and Sutton, B. C. (1974) Article 59 and names of perfect state taxa in imperfect state genera. *Taxon* **23**: in press.

Hawksworth, D. L. and Wells, H. (1973) Ornamentation on the terminal hairs in *Chaetomium* Kunze ex Fr. and allied genera. *Mycol. Pap.* **134**: 1–24.

Hawksworth, F. G. and Wiens, D. (1972) Biology and classification of Dwarf Mistletoes (*Arceuthobium*). *U.S. Dept. Agric. Forest Serv., Agric. Handb.* **401**: i–viii, 1–234.

Hayat, M. A. (1970) *Principles and Techniques of Electron Microscopy 1, Biological Applications.* New York, etc.: Van Nostrand, Reinhold.

Hearle, J. W. S., Sparrow, J. T. and Cross, P. M. (1972) *The Use of the Scanning Electron Microscope.* Oxford and New York: Pergamon Press.

Heath, J. and Scott, D. (1972) *Instructions for Recorders.* Huntingdon: Nature Conservancy Biological Records Centre.

Hedge, I. C. and Lamond, J. M. (1970) *Index of Collectors in the Edinburgh Herbarium.* Edinburgh: H.M.S.O.

Henderson, D. M., Orton, P. D. and Watling, R. (1969) *British Fungus Flora, Agarics and Boleti, Introduction.* Edinburgh: H.M.S.O.

Henssen, A. (1970) Die Apothecienentwicklung bei *Umbilicaria* Hoffm. emend. Frey. *Votr. bot. Ges.* [*Dt. bot. Ges.*] N.F. **4**: 103–126.

Henssen, A. and Jahns, H. M. (1974) *Lichenes.* Stuttgart: Thieme.

Heywood, V. H. (1958) *The Presentation of Taxonomic Information.* Leicester: Leicester University Press.

Heywood, V. H. (1959) The taxonomic treatment of ecotypic variation. *In* Cain, A. J. (ed.) *Function and Taxonomic Importance*: 87–112. London: Systematics Association.

Heywood, V. H. (1963) The 'species aggregate' in theory and practice. *Regnum Vegetabile* **27**: 26–37.

Hill, T. G. (1915) *The Essentials of Illustration.* London: Wesley.

Hodgkiss, A. G. (1970) *Maps for Books and Theses.* Newton Abbott: David and Charles.

Höfler, K. (1938) Pilzsoziologie. *Ber. dt. bot. Ges.* **55**: 606–622.

Höhnel, F. (1909) Fragmente zur Mykologie (VI. Mitteilung, Nr. 182 bis 288). *Sber. Akad. wiss. Wien, mat.-nat. Kl., Abt.* I, **118**: 275–452.

Holden, M. (ed.) (1969) Guide to the literature for the identification of British fungi. *Bull. Br. mycol. Soc.* **3**: 19–54.

Hooker, J. D. and Jackson, B. D. (1895 →) *Index Kewensis.* 2 vols. and *Supplements* **1** →. Oxford: Clarendon Press.

Howe, R. H. (1911) The lichens of the Linnean herbarium with remarks on Acharian material. *Bull. Torrey Bot. Club* **39**: 199–203.

Hudson, H. J. (1970) Infraspecific categories in fungi. *Biol. J. Linn. Soc.* **2**: 211–219.

Hueck, H. J. (1953) Myco-sociological methods of investigation. *Vegetatio* **4**: 84–101.

Huneck, S. (1968) Lichen substances. *In* Reinhold, L. and Liwschitz, Y. (ed.) *Progress in Phytochemistry* **1**: 223–346. London, etc.: Wiley Interscience.

Huneck, S. (1971) Chemie und Biosynthese der Flechtenstoffe. *Fortschr. Chem. org. Naturst.* **29**: 209–306.

Huneck, S. and Linscheid, P. (1968) 45. Mitteilung über Flechteninhaltstoffe, NMR-Spektroskopie einiger Depside und Depsidone. *Z. Naturf.* **23b**: 717–732.

Huxley, J. S. (1940) Towards the new systematics. *In* Huxley, J. [S.] (ed.) *The New Systematics*: 1–46. London: Oxford University Press.

Iizuka, H. and Hasegawa, T. (ed.) (1970) *Culture Collections of Microorganisms.* Baltimore: University Park Press.

Ingold, C. T. (1971) *Fungal Spores, their liberation and dispersal.* Oxford: Clarendon Press.

Isaac, I. and Davies, R. R. (1955) A new hyaline species of *Verticillium: V. intertextum* sp. nov. *Trans. Br. mycol. Soc.* **38**: 143–156.

Ivimey-Cook, R. B. (1969) Investigations into the phenetic relationships between species of *Ononis* L. *Watsonia* **7**: 1–23.

Jackson, B. D. (1928) *A Glossary of Botanic Terms.* Ed. 4. London: Duckworth.

Jacobs, J. B. and Ahmadjian, V. (1969) The ultrastructure of lichens. I. A general survey. *J. Phycol.* **5**: 227–240.

Jaeger, E. C. (1955) *A Source Book of Biological Names and Terms.* Ed. 3. Springfield, Ill.: Thomas.

Jahns, H. M. (1970) Untersuchungen zur Entwicklungsgeschichte der Cladoniaceen. *Nova Hedwigia* **20**: 1–177.

Jahns, H. M. and Beltman, H. A. (1973) Variations in the ontogeny of fruiting bodies in the genus *Cladonia* and their taxonomic and phylogenetic significance. *Lichenologist* **5**: 349–367.

James, P. W. (1971) New or interesting British lichens: 1. *Lichenologist* **5**: 114–148.

Janex-Favre, M. C. (1971) Recherches sur l'ontogénie, l'organisation et les asques de quelques Pyrénolichens. *Revue bryol. lichén.* **37**: 421–650.

Jefferys, E. G. (1972) A scheme for the numerical classification of fungi. *Bull. Br. mycol. Soc.* **6**: 25–28.

Jeffrey, C. (1973) *Biological Nomenclature.* London: Arnold.

Jinks, J. L. and Croft, J. (1971) Methods used for genetical studies in mycology. *In* Booth, C. (ed.) *Methods in Microbiology* **4**: 479–500. London and New York: Academic Press.

Johnson, T. (1968) Host specialization as a taxonomic criterion. *In* Ainsworth, G. C.

and Sussman, A. S. (ed.) *The Fungi, An advanced treatise* **3**: 543–556. New York and London: Academic Press.

Jones, E. B. G. (1971) Aquatic fungi. *In* Booth, C. (ed.) *Methods in Microbiology* **4**: 335–365. London and New York: Academic Press.

Kapp, E. (1959) Les collections de l'Institut de Botanique de la Faculté des Sciences de Strasbourg. *Bull. Soc. bot. Fr.* **106**: 197–198.

Kärenlampi, L. and Pelkonen, M. (1971) Studies on the morphological variation of the lichen *Cladonia uncialis*. *Rept Kevo Subarct. Res. Stn* **7**: 47–56.

Kendrick, W. B. (1969) Preservation of fleshy fungi for taxonomy. *Mycologia* **61**: 392–395.

Kendrick, [W.] B. (ed.) (1971) *Taxonomy of Fungi Imperfecti*. Toronto: University of Toronto Press.

Kendrick, W. B. (1972) Computer graphics in fungal identification. *Can. J. Bot.* **50**: 2171–2175.

Kent, D. H. (1957) *British Herbaria*. London: Botanical Society of the British Isles.

Kershaw, K. A. (1964) *Quantitative and Dynamic Ecology*. London: Arnold.

Kershaw, K. A. and Millbank, J. W. (1969) A controlled environment lichen growth chamber. *Lichenologist* **4**: 83–87.

Kershaw, K. A. and Millbank, J. W. (1970) Isidia as vegetative propagules in *Peltigera aphthosa* var. *variolosa* (Massal.) Thoms. *Lichenologist* **4**: 214–217.

Klement, O. (1955) Prodromus der mitteleuropäischen Flechtengesellschaften. *Reprium nov. Spec. Regni veg., Beih.* **135**: 5–194.

Klement, O. (1960) Zur Flechtenvegetation der Oberpfalz. *Ber. Bayer. bot. Ges.* **28**: 250–275.

Koehler, J. K. (ed.) (1973) *Advanced Techniques in Biological Electron Microscopy*. Berlin, etc.: Springer.

Kohlmeyer, J. (1962a) Index alphabeticus Klotzschii et Rabenhorstii Herbarii Mycologici. *Beih. Nova Hedwigia* **4**: i–xvi, 1–231.

Kohlmeyer, J. (1962b) Die Pilzsammlung des Botanischen Museums zu Berlin Dahlem (B). *Willdenowia* **3**: 63–70.

Korf, R. P. (1972) Synoptic key to the genera of the Pezizales. *Mycologia* **64**: 937–994.

Kornerup, A. and Wanscher, J. H. (1967) *Methuen Handbook of Colour*. Ed. 2. London: Methuen.

Krauss, H. M. (1973) The use of generalized information processing systems in the biological sciences. *Taxon* **22**: 3–18.

Krempelhuber, A. von (1867, 1869, 1872) *Geschichte und Litteratur der Lichenologie*. 3 vols. Munich: Wolf.

Krieger, L. C. C. (1924) *Catalogue of the Mycological Library of Howard A. Kelly*. Baltimore, Mass.: Privately printed.

Kylin, H. (1954) *Die Gattungen der Rhodophyceen*. Lund: Gleerup.

Lamb, I. M. (1963) *Index nominum lichenum inter annos 1932 et 1960 divulgatorum*. New York: Ronald Press.

Lamb, I. M. (1968) Antarctic lichens II. The genera *Buellia* and *Rinodina*. *Br. Antarct. Surv. Sci. Rept* **61**: 1–129.

Lange, M. (1968) Genetical and cytological aspects of taxonomy. *In* Ainsworth, G. C. and Sussman, A. S. (ed.) *The Fungi, An advanced treatise* **3**: 625–631. New York and London: Academic Press.

Lanjouw, J. (1958) On the nomenclature of chemical strains. *Taxon* **7**: 43–44.

Lanjouw, J. and Stafleu, F. A. (1954) Index Herbariorum Part II, Collectors A–D. *Regnum Vegetabile* **2**: 1–174.

Lanjouw, J. and Stafleu, F. A. (1956) Index Herbariorum Part II(2), Collectors E–H. *Regnum Vegetabile* **9**: 175–295.

Lanjouw, J. et al. (ed.) (1966) International Code of Botanical Nomenclature adopted by the Tenth International Botanical Congress Edinburgh, August 1964. *Regnum Vegetabile* **46**: 1–402.

Laundon, G. F. (1968) Microscopy. *In* Commonwealth Mycological Institute, *Plant Pathologist's Pocketbook*: 224–226. Kew: Commonwealth Mycological Institute.

Laundon, G. F. (1971) A new reinforcement for sealed fluid microslide mounts. *Trans. Br. mycol. Soc.* **56**: 317–318.

Laundon, J. R. (1967) A study of the lichen flora of London. *Lichenologist* **3**: 277–327.

Laundon, J. R. (1970) Lichens new to the British flora: 4. *Lichenologist* **4**: 297–308.
Lawrence, G. H. M., Buchheim, A. F. G., Daniels, G. S. and Dolezal, D. (1968) *Botanico-Periodicum Huntianum*. Pittsburgh: Hunt Botanical Library.
Lawson, D. F. (1972) *Photomicrography*. London and New York: Academic Press.
Leach, C. M. (1971) A practical guide to the effects of visible and ultraviolet light on fungi. *In* Booth, C. (ed.) *Methods in Microbiology* **4**: 609–664. London and New York: Academic Press.
Leenhouts, P. W. (1966a) Keys in biology. I. A survey and a proposal of a new kind. *Proc. K. ned. Akad. Wet., ser. C*, **69**: 571–586.
Leenhouts, P. W. (1966b) Keys in biology. II. A survey and a proposal of a new kind. *Proc. K. ned. Akad. Wet., ser. C*, **69**: 587–596.
Leenhouts, P. W. (1968) A guide to the practice of herbarium taxonomy. *Regnum Vegetabile* **58**: 1–60.
Lellinger, D. B. (1972) Dichlorvos and lindane as herbarium insecticides. *Taxon* **21**: 91–95.
Lentz, M. E. and Lentz, P. L. (1968) The National Fungus Collection. *Bioscience* **18**: 194–200.
Lentz, P. L. and Hawksworth, D. L. (1971) Typification of *Ascotricha* species described by L. M. Ames. *Trans. Br. mycol. Soc.* **57**: 317–324.
Letrouit-Galinou, M.-A. (1968) The apothecia of the discolichens. *Bryologist* **71**: 297–327.
Letrouit-Galinou, M.-A. (1971) Études sur le '*Lobaria laetevirens*' (Lght.) Zahlb. *Botaniste* **54**: 189–234.
Letrouit-Galinou, M.-A. and Lallement, R. (1971) Le thalle, les apothécies et les asques du *Peltigera rufescens* (Weis) Humb. (Discolichen, Peltigeracée). *Lichenologist* **5**: 59–88.
Lewis, C. T. and Short, C. (1955) *A Latin Dictionary*. Revised Ed. Oxford: Clarendon Press.
Lindau, G. and Sydow, P. (1908–18) *Thesaurus litteraturae mycologicae et lichenologicae*. 5 vols. Berlin: Borntraeger.
Linnaeus, C. (1753) *Species plantarum*. 2 vols. Stockholm.
Lisiewska, M. (1972) Mycosociological research on macromycetes in beech forest associations. *Mycopath. Mycol. appl.* **48**: 23–34.
Lockhart, W. R. and Liston, J. (ed.) (1970) *Methods for Numerical Taxonomy*. Bethesda, Maryland: American Society for Microbiology.
Lundqvist, N. (1972) Nordic Sordariaceae s. lat. *Symb. bot. upsal.* **20**(1): 1–374.
McVaugh, R., Ross, R. and Stafleu, F. A. (1968) An annotated glossary of botanical nomenclature. *Regnum Vegetabile* **56**: 1–31.
McVean, D. N. and Ratcliffe, D. A. (1962) Plant communities of the Scottish Highlands. *Monogr. Nature Conservancy* **1**: i–xiii, 1–445. London: H.M.S.O.
Magnusson, A. H. (1940) Studies in species of *Pseudocyphellaria*—The *Crocata*-group. *Acta Horti Gotob.* **14**: 1–36.
Majewski, T. (1971) Grzyby pasożytnicze Białowieskiego Parku Narodowego na tle mikoflory Polski. *Acta mycol.* **7**: 299–388.
Martin, J. (1968) O primenenii perfokart pri opredelenii lishaĭnikov. *Trans. Tartu St. Univ.* **211**, *Pap. Bot.* **8**: 130–135.
Martin, S. M. and Skerman, V. B. D. (ed.) (1972) *World Directory of Collections of Cultures of Microorganisms*. New York, etc.: Wiley Interscience.
Martinsen, W. L. M. (1968) Photography of specimens. *In* Engel, C. E. (ed.) *Photography for the scientist:* 451–482. London and New York: Academic Press.
Mason, E. W. (1940) On specimens, species and names. *Trans. Br. mycol. Soc.* **24**: 115–125.
Massé, L. J. C. (1966) Flore et végétation lichéniques des Iles Glénan (Finistère). *Revue bryol. lichén.* **34**: 854–927.
Mayr, E., Linsley, E. F. and Usinger, R. L. (1953) *Methods and Principles of Systematic Zoology*. New York: McGraw-Hill.
Meijer Drees, E. (1953) A tentative design for rules of phytosociological nomenclature. *Vegetatio* **3**: 205–214.
Merriam, G., Merriam, C. and Co. (1972) *Webster's Geographical Dictionary*. Revised Ed. Springfield, Mass.: Merriam.
Metcalf, Z. P. (1954) The construction of keys. *Syst. Zool.* **3**: 38–45.

Mills, F. W. (1933–35) *An Index to the Genera of the Diatomaceae.* London: Wheldon and Wesley.

Morse, L. E. (1968) Construction of identification keys by computer. *Am. J. Bot.* **55**: 737.

Muséum National d'Histoire Naturelle (1954) *La Chaire de Cryptogamie du Muséum National d'Histoire Naturelle, Organisation et Buts.* Paris: Muséum National d'Histoire Naturelle.

Muzeelor, R. (ed.) (1971) *Directory of the Natural Science Museums of the World.* Bucurest: Conseil International des Musées.

Neelakantan, S. and Rao, P. S. (1967) Place of chemistry in lichen taxonomy. *Bull. natn. Inst. Sci. India* **34**: 168–178.

Nissen, C. (1966) *Die botanische Buchillustration.* Stuttgart: Hiersmann.

Nobles, M. K. (1965) Identification of cultures of wood-inhabiting Hymenomycetes. *Can. J. Bot.* **43**: 1097–1139.

Norris, J. R. (1968) The application of gel electrophoresis to the classification of micro-organisms. *In* Hawkes, J. G. (ed.) *Chemotaxonomy and Serotaxonomy*: 49–56. London and New York: Academic Press.

Oatley, C. W. (1972) *The Scanning Electron Microscope, Part 1 The Instrument.* London: Cambridge University Press.

Olá'h, G.-M. (1970) Le genre *Panaeolus. Rev. mycol.*, *Mém.* **10**: 1–273.

Olá'h, G.-M. (1972) L'ontogénie sporale et l'ultrastructure de la paroi chez *Paecilomyces berlinensis. Can. J. Microbiol.* **18**: 1471–1475.

Oliver, K. D. (1965) *Catalogue of the Buller Memorial Library.* Ottawa: Canada Department of Agriculture.

Onions, A. H. S. (1971) Preservation of fungi. *In* Booth, C. (ed.) *Methods in Microbiology* **4**: 113–151. London and New York: Academic Press.

Osborne, D. V. (1963) Some aspects of the theory of dichotomous keys. *New Phytol.* **62**: 144–160.

Pankhurst, R. J. (1970) Key generation by computer. *Nature, Lond.* **227**: 1269–1270.

Pankhurst, R. J. (1971) Botanical keys generated by computer. *Watsonia* **8**: 357–368.

Pankhurst, R. J. and Walters, S. M. (1971) Generation of keys by computer. *In* Cutbill, J. L. (ed.) *Data Processing in Biology and Geology*: 189–203. London and New York: Academic Press.

Parker-Rhodes, A. F. and Jackson, D. M. (1969) Automatic classification in the ecology of the higher fungi. *In* Cole, A. J. (ed.) *Numerical Taxonomy*: 180–215. London and New York: Academic Press.

Parkinson, D. and Williams, S. T. (1961) A method for isolating fungi from soil microhabitats. *Plant Soil* **13**: 347–355.

Parmelee, J. A. (1971) Models of fungi for public display. *Greenhouse-Garden-Grass* **10**: 73–77.

Pearson, L. C. (1970) Varying environmental factors in order to grow intact lichens under laboratory conditions. *Am. J. Bot.* **75**: 659–664.

Pegler, D. N. and Young, T. W. (1970) Basidiospore morphology in the Agaricales. *Beih. Nova Hedwigia* **35**: 1–210.

Pelletier, G. and Hall, R. (1971) Relationships among species of *Verticillium*: protein composition of spores and mycelium. *Can. J. Bot.* **49**: 1293–1297.

Perring, F. [H.] (1971) The British biological recording network. *In* Cutbill, J. L. (ed.) *Data Processing in Biology and Geology*: 115–121. London and New York: Academic Press.

Perring, F. H. and Walters, S. M. (1962) *Atlas of the British Flora.* London: Nelson.

Petrak, F. (1930–44) Verzeichnis der neuen Arten, Varietäten, Formen, Namen und wichtigsten Synonyme. *Just's bot. Jber.* **48**(3): 184–256; **49**(2): 267–336; **56**(2): 291–697; **57**(2): 592–631; **58**(1): 447–570; **60**(1): 449–514; **63**(2): 805–1056.

Petrak, F. (1950) *Index of Fungi 1936–1939.* Kew: Commonwealth Mycological Institute.

Pilát, A. (1938) Seznam druhů hub, popsaných A. C. J. Cordou s udáním originálních exemplářů, které jsou uloženy v herbáři Národní ho Musea v Praze. *Sbornik Národ. Mus. Praze* **1B**: 139–170.

Plumb, R. T. and Turner, R. H. (1972) Scanning electron microscopy of *Erysiphe graminis. Trans. Br. mycol. Soc.* **59**: 149–178.

Poelt, J. (1972) Die taxonomische Behandlung von Artenpaare bei den Flechten. *Bot. Notiser* **125**: 77–81.

Potztal, E. (1961) Bericht über den Botanischen Garten und das Botanische Museum zu Berlin-Dahlem für das Jahr 1960. *Willdenowia* 2: 775–799.

Preece, T. F. (1968) The fluorescent antibody technique for the identification of plant pathogenic fungi. *In* Hawkes, J. G. (ed.) *Chemotaxonomy and Serotaxonomy*: 111–114. London and New York: Academic Press.

Preece, T. F. (1971a) Immunological techniques in mycology. *In* Booth, C. (ed.) *Methods in Microbiology* 4: 599–607. London and New York: Academic Press.

Preece, T. F. (1971b) Fluorescent techniques in mycology. *In* Booth, C. (ed.) *Methods in Microbiology* 4: 509–516. London and New York: Academic Press.

Proctor, A. G. (1967) Serological methods in mycology. *In* Collins, C. H. (ed.) *Progress in Microbiological Methods*: 213–226. London: Butterworths.

Pritzel, G. A. (1871–77) *Thesaurus Literaturae botanicae*. Ed. 2. Leipzig: Brockhaus.

Punithalingam, E. (1972) Cytology of *Fusarium culmorum*. *Trans. Br. mycol. Soc.* 58: 225–230.

Purseglove, J. W. (1968, 1972) *Tropical Crops*. 4 vols. London: Longman.

Rao, V. R. and Mukerji, K. G. (1970) Cytology of the ascus in *Ascotricha guamensis*. *Mycologia* 62: 301–306.

Raper, K. B. and Thom, C. (1949) *A Manual of the Penicillia*. Baltimore: Williams and Wilkins.

Raphael, S. (1970) The publication dates of the *Transactions of the Linnean Society of London*, series I, 1791–1875. *Biol. J. Linn. Soc.* 2: 61–76.

Rayner, R. W. (1970) *A Mycological Colour Chart*. Kew: Commonwealth Mycological Institute.

Rayner, R. W. (1973) Cards for recording fungi. *Bull. Br. mycol. Soc.* 7: 15–24.

Reddy, M. N. and Stathmann, M. A. (1972) Isozyme patterns of *Fusarium* species and their significance in taxonomy. *Phytopath. Z.* 74: 115–125.

Reid, D. A. (1966→) Coloured icones of rare and interesting fungi. Part I→. *Nova Hedwigia* 11→. [As Supplements.]

Richards, M. (1972) Serology and yeast classification. *Antonie van Leeuwenhoek* 38: 177–192.

Richardson, D. H. S. (1967) Transplantation of lichen thalli to solve some taxonomic problems in *Xanthoria parietina* (L.) Th. Fr. *Lichenologist* 3: 386–391.

Richardson, D. H. S. (1971) Lichens. *In* Booth, C. (ed.) *Methods in Microbiology* 4: 267–293. London and New York: Academic Press.

Ridgway, R. (1912) *Colour Standards and Nomenclature*. Washington: Ridgway.

Robinson, R. A. (1969) Disease resistance terminology. *Rev. appl. Mycol.* 48: 593–606.

Romagnesi, H. (1970) *Nouvelle Atlas des Champignons*. 4 vols. Ed. 2. Paris: Bordas.

Rose, G. G. (ed.) (1963) *Cinemicrography in Cell Biology*. London and New York: Academic Press.

Royal Geographical Society (1954) *The Cartographical Presentation of Biological Distributions*. London: Royal Geographical Society.

Royal Society (1965) *General notes on the Preparation of Scientific Papers*. Ed. 2. London: Royal Society.

Royal Society (1971) *Quantities, Units, and Symbols*. London: Royal Society.

Saccardo, P. A. (1882–1931, 1972) *Sylloge Fungorum*. 26 vols. Pavia: P. A. Saccardo; New York: Johnson.

Santesson, J. (1967) Chemical studies on lichens. 4. Thin layer chromatography of lichen substances. *Acta Chem. Scand.* 21: 1162–1172.

Santesson, R. (1952) Foliicolous lichens I, A revision of the taxonomy of the obligately foliicolous, lichenized fungi. *Symb. bot. upsal.* 12(1): 1–590.

Santesson, R. (1968) Lavar. Some aspects on lichen taxonomy. *Svensk Natur.* 1968: 176–184.

Savage, S. (1945) *A Catalogue of the Linnean Herbarium*. London: Taylor and Francis.

Savile, D. B. O. (1962) *Collection and Care of Botanical Specimens*. Ottawa: Canada Department of Agriculture.

Sayre, G. (1969) Cryptogamae Exsiccatae—An annotated bibliography of published exsiccatae of Algae, Lichenes, Hepaticae, and Musci. *Mem. N.Y. Bot. Gdn* 19: 1–174.

Sayre, G., Bonner, C. E. B. and Culberson, W. L. (1964) The authorities for the epithets of mosses, hepatics, and lichens. *Bryologist* 67: 113–135.

Scott, P. J. (1973) A consideration of the category in classification. *Taxon* **22**: 405–406.

Sheard, J. W. (1962) A contribution to the lichen flora of Jan Mayen. *Lichenologist* **2**: 76–85.

Sheard, J. W. (1973) *Rinodina interpolata* (Stirt.) Sheard, a new combination in the British and Scandinavian lichen floras. *Lichenologist* **5**: 461–463.

Shelter, S. G., Beaman, J. H., Hale, M. E., Morse, L. E., Crockett, J. J. and Creighton, R. A. (1971) Pilot data processing systems for floristic information. *In* Cutbill, J. L. (ed.) *Data Processing in Biology and Geology*: 275–310. London and New York: Academic Press.

Shelter, S. G. and Read, R. W. (ed.) (1973) International index of current research projects in plant systematics. *Flora N. Am. Rept* **71**: i-xxii, 1–118.

Shibata, S., Natori, S. and Udagawa, S. (1964) *List of Fungal Products*. Tokyo: University of Tokyo Press.

Shimwell, D. W. (1971) *The Description and Classification of Vegetation*. London: Sidgwick and Jackson.

Simpson, N. D. (1960) *A Bibliographical Index of the British Flora*. Bournemouth: Privately printed.

Smith, A. L. (1921) *Lichens*. Cambridge: Cambridge University Press.

Smith, C. E. (1971) Preparing herbarium specimens of vascular plants. *U.S. Dept. Agric., Agric. Inf. Bull.* **348**: i-vi, 1–29.

Smith, I. (1960) *Chromatographic and Electrophoretic Techniques*. 2 vols. London: Heinemann; New York: Wiley Interscience.

Smith, R. S. (1967) Control of tarsonemial mites in fungal cultures. *Mycologia* **59**: 600–609.

Sneath, P. H. A. (1971) Numerical taxonomy: criticisms and critiques. *Biol. J. Linn. Soc.* **3**: 147–157.

Sneath, P. H. A. and Sokal, R. R. (1973) *Numerical Taxonomy*. San Francisco: Freeman.

Snell, W. H. and Dick, E. A. (1971) *A Glossary of Mycology*. Revised Ed. Cambridge, Mass.: Harvard University Press.

Sokal, R. R. and Sneath, P. H. A. (1963) *Principles of Numerical Taxonomy*. San Francisco: Freeman.

Soper, J. H. (1969) The use of data processing methods in the herbarium. *Ann. Inst. Biol. Univ. Nal. Antón. México, ser. Bot.* **40**: 105–116.

Stafleu, F. A. (1967) Taxonomic literature, A selective guide to botanical publications with dates, commentaries and types. *Regnum Vegetabile* **52**: i-xx, 1–556.

Stafleu, F. A. (1972) The volumes on cryptogams of 'Engler und Prantl'. *Taxon* **21**: 501–511.

Stafleu, F. A. *et al.* (ed.) (1972) International Code of Botanical Nomenclature adopted by the Eleventh International Botanical Congress, Seattle, August 1969. *Regnum Vegetabile* **82**: 1–426.

Staniland, L. N. (1952) *The Principles of Line Illustration*. London: Burke.

Stearn, W. T. (1951) Mapping the distribution of species. *In* Lousley, J. E. (ed.) *The Study of the Distribution of British Plants*: 48–64. London: Botanical Society of the British Isles.

Stearn, W. T. (1956) Keys, botanical, and how to use them. *In* Synge, P. M. (ed.) *Dictionary of Gardening, Supplement*: 251–253. Oxford: Clarendon Press.

Stearn, W. T. (1959) The background of Linnaeus's contributions to the nomenclature and methods of systematic biology. *Syst. Zool.* **8**: 4–22.

Stearn, W. T. (1968) Observations on a computer-aided survey of the Jamaican species of *Columnea* and *Alloplectus*. *In* Heywood, V. H. (ed.) *Modern Methods in Plant Taxonomy*: 219–223. London and New York: Academic Press.

Stearn, W. T. (1971) Sources of information about botanic gardens and herbaria. *Biol. J. Linn. Soc.* **3**: 225–233.

Stearn, W. T. (1973) *Botanical Latin*. Ed. 2. Newton Abbot: David and Charles.

Stebbins, G. L. (1950) *Variation and Evolution in Plants*. New York and London: Columbia University Press.

Stevenson, J. A. (1960) A list of authors of plant parasite names with recommended abbreviations. *In* U.S. Department of Agriculture, Index of plant diseases in the United States. *U.S. Dept. Agric., Agric. Handb.* **165**: 517–547.

Stevenson, J. A. (1967) Rabenhorst and Fungus exsiccati. *Taxon* **16**: 112–119.

Stevenson, J. A. (1971) An account of fungus exsiccati containing material from the Americas. *Beih. Nova Hedwigia* **36**: i-vii, 1–563.

Stockdale, P. M. (1971) Fungi pathogenic for man and animals: I. Diseases of the keratinized tissues. *In* Booth, C. (ed.) *Methods in Microbiology* **4**: 429–460. London and New York: Academic Press.

Sutton, B. C. (1971) Coelomycetes. IV. The genus *Harknessia*, and similar fungi on *Eucalyptus*. *Mycol. Pap.* **123**: 1–46.

Sutton, B. C. and Sandhu, D. K. (1969) Electron microscopy of conidium development and secession in *Cryptosporiopsis* sp., *Phoma fumosa*, *Melanconium bicolor*, and *M. apiocarpum*. *Can. J. Bot.* **47**: 745–749.

Swartz, D.† (1971). *Collegiate Dictionary of Botany*. New York: Ronald Press.

Swinscow, T. D. V. (1962) Pyrenocarpous lichens: 3 The genus *Porina* in the British Isles. *Lichenologist* **2**: 6–56.

Systematics Association Committee for Descriptive Biological Terminology (1960) 1. Preliminary list of works relevant to descriptive biological terminology. *Taxon* **9**: 245–257.

Systematics Association Committee for Descriptive Biological Terminology (1962) II. Terminology of simple symmetrical plane shapes (Chart 1). *Taxon* **11**: 145–156.

Talbot, P. H. B. (1971) *Principles of Fungal Taxonomy*. London and Basingstoke: Macmillan Press.

Tarr, S. A. J. (1972) *Principles of Plant Pathology*. London and Basingstoke: Macmillan Press.

Tétényi, P. (1970) *Infraspecific Chemical Taxa of Medicinal Plants*. Budapest: Akadémial Kiadó.

Thoen, D. (1970) Étude mycosociologique de quelques associations forestières des districts Picardo-Brabançon, Mosau et Ardennais de Belgique. *Bull. agron. Gembloux*, *n.s.* **5**: 309–326.

Thomson, J. W. (1968) ['*1967*'] *The Lichen Genus* Cladonia *in North America*. Toronto: University of Toronto Press.

Tibell, L. (1971) The genus *Cyphelium* in Europe. *Svensk bot. Tidskr.* **65**: 138–164.

Tribe, H. T. (1972) Sealing of lactophenol mounts. *Trans. Br. mycol. Soc.* **58**: 341.

Tulloch, M. (1972) The genus *Myrothecium* Tode ex Fr. *Mycol. Pap.* **130**: 1–42.

Tupholme, C. H. S. (1961) *Colour Photomicrography with a 35 mm Camera*. London: Faber and Faber.

Turner, R. P. (1964) *Technical Writer's and Editor's Stylebook*. Indianapolis: Sams.

Turner, W. B. (1971) *Fungal Metabolites*. London and New York: Academic Press.

Turrill, W. B. (1964) Floras. *Vistas in Botany* **4**: 225–238.

Ubrizsy, G. (1972) Nouvelles analyses mycocénologiques dans certains types sylvicoles de la Hongrie. *Mycopath. Mycol. appl.* **48**: 75–86.

Uphof, J. C. T. (1968) *Dictionary of Economic Plants*. Lehre: Cramer.

Uyenco, F. R. (1965) Studies on some lichenized *Trentepohlia* associated in lichen thalli with *Coenogonium*. *Trans. Am. microsc. Soc.* **84**: 1–14.

Van Beverwijk, A. L. (1961) Are 'type cultures' type material? *In* Commonwealth Mycological Institute (ed.) *Report of the Sixth Commonwealth Mycological Conference*: 33–35. Kew: Commonwealth Mycological Institute.

Van der Wijk, R., Margadant, W. D. and Florschütz, P. A. (1959–69) *Index Muscorum* I–V. *Regnum Vegetabile* **17**: 1–548; **26**: 1–535; **33**: 1–529; **48**: 1–604; **65**: i-xii, 1–922.

Vainio, E. A. (1886a) Revisio lichenum in herbario Linnaei asservatorum. *Medd. Soc. Fl. Fauna fenn.* **14**: 9–10.

Vainio, E. A. (1886b) Revisio lichenum Hoffmannianorum. *Medd. Soc. Fl. Fauna fenn.* **14**: 11–19.

Vanlandingham, S. L. (1967) *Catalogue of the Fossil and Recent Genera and Species of Diatoms and their Synonyms*. Lehre: Cramer.

Van Steenis-Kruseman, M. J. and Stearn, W. T. (1954) Dates of publication. *Fl. Malesiana*, *ser*. 1, **4**(5): clxiii-ccxix.

Verseghy, K. (1964) *Typen-Verzeichnis der Flechtensammlung in der Botanisches Abteilung des Ungarisches Naturwissenschaftlichen Museums*. Budapest: Természettudományi Múzeum.

Verseghy, K. (1968) Nachtrag I. zum 'Typen-Verzeichnis der Flechtensammlung in der

Botanisches Abteilung des Ungarisches Naturwissenschaftlichen Museums' zusammengestellt. *Fragm. Bot. Mus. Hist.-Nat. Hungar.* **6**: 41–55.

Viégas, A. P. (1961) *Índice de fungos da América do Sul.* Campinas: Instituto Agronômico.

Wade, A. E. (1959) Soil binder for crustaceous ground lichens. *Lichenologist* **1**: 87–88.

Wanstall, P. J. (ed.) (1963) *Local Floras.* London: Botanical Society of the British Isles.

Ware, W. M. (1933) A disease of cultivated mushrooms caused by *Verticillium malthousei* sp.nov. *Ann. Bot.* **47**: 763–785.

Watling, R. (1970) Spore colour. *Bull. Br. mycol. Soc.* **4**: 48–49.

Watling, R. (1971) Chemical tests in Agaricology. *In* Booth, C. (ed.) *Methods in Microbiology* **4**: 567–597. London and New York: Academic Press.

Weakley, B. S. (1972) *A Beginner's Handbook in Biological Electron Microscopy.* Edinburgh and London: Livingstone.

Weber, W. A. (1968) A taxonomic revision of *Acarospora*, subgenus *Xanthothallia*. *Lichenologist* **4**: 16–31.

Webster, J. (1970) *Introduction to Fungi.* Cambridge: Cambridge University Press.

Wheeler, B. E. J. (1968) Fungal parasites of plants. *In* Ainsworth, G. C. and Sussman, A. S. (ed.) *The Fungi, An advanced treatise* **3**: 179–210. New York and London: Academic Press.

Williams, P. C. (1968) *Abbreviated Titles of Biological Journals.* Ed. 3. London: Biological Council.

Willis, J. C. (1973) *A Dictionary of the Flowering Plants and Ferns.* Ed. 8 (revised H. K. Airey-Shaw). Cambridge: Cambridge University Press.

Wirth, V. (1972) Die Silikatflechten-Gemeinschaften im außeralpinen Zentraleuropa. *Diss. Bot., Lehre* **17**: 1–306, *1–9.*

Woods, R. S. (1966) *An English-Classical Dictionary for the use of Taxonomists.* Covina: Pomona College.

Wright, J. E. and Calonge, F. D. (1973) The location of Lazaro e Ibiza's collections of Polyporaceae. *Taxon* **22**: 267–270.

Wright, J. E. and Lois, R. J. (1949) Lista de siglas de autores empleadas en micologia y fitopatologia. *Lilloa* **21**: 225–269.

Yarranton, G. S. (1967) A quantitative study of the bryophyte and macrolichen vegetation of the Dartmoor granite. *Lichenologist* **3**: 392–408.

Zahlbruckner, A. (1921–40) *Catalogus lichenum universalis.* 10 vols. Leipzig: Borntraeger.

Zoberi, M. H. (1972) *Tropical Macrofungi, some common species.* London and Basingstoke: Macmillan Press.

X. INDEX

Pages including definitions of terms are indicated in *italic* type

synonym(s, -um), 63, 66, 79, 110, 126, 129–30, 140, 153, 155, 162, *206*
  absolute, *200*
  earliest, *201*
  facultative, 130, *202*
  heterotypic, 130, *202*
  homotypic, 130, *202*
  later, *202*
  nomenclatural, 130, 140, *204*
  objective, *204*
  obligate, 130, *204*
  older, *204*
  replaced, 153–4
  subjective, *205*
  taxonomic, 130, 140, *206*
  younger, 130, *207*
synonymy, 66–8, 162–3, *206*
  layout of, 66–8
synoptical key, *74*
syntaxa, *see* names of communities
synopsis, 117; *see also* abstract
syntype, 127–8, *135*, *206*

tables, 62
tautonym, *147*, 164, *206*
taxa, *see* taxon
taxometrics, 102; *see also* numerical taxonomy
taxon, *133*, *206*
  vag(um), *206*
taxonomic ranks, 38–47, 133
  synonym, 130, 140, *206*
taxonomists, types of, 48
term 144, *206*; *see also* glossaries
  descriptive, 52–4
  misplaced, *154*, *203*
termination, 39, 42, 90–2, 142–5, 147–8, 152, 168, 173–7, *206*
teste, *206*
thin-layer chromatography, 24–5, 98–9
time-lapse photomicrography, 61, 104
title abbreviations, books, 67, 193–9
  journals, 67, 110, 114, 193, 198
  selection, 117
  style, 42
topotype, *206*
transfer, *206*
transplants, 34, 105
tribe, 38–40, 133, 143–4, 154
tropical microfungi, collecting, 17
TWA, *34*
type, 132, *135*, *206*; *see also* typification
  conservation of, 130, 136, 161–3

culture, 36, 65, 101, 127, 137, *206*
  nomenclatural, 66, 127, *135*, *204*
  record, 92–3, 97
  specimens, 63–5, 67–8, 127, *206*
    citation, 67–8
    location, 179–92
    preservation, 26, 29, 31–2, 64, 127, 137, 179
  typification, 39, 63, 65–6, 125–30, 132, 134–8, 141–2, 147, 156, 158, 161, 163–4, 166–7, 171–2, 178, *206*
    automatic, 135
    communities, 92–3
  typographic error, 41, 173, *206*
  typonym, 130, *206*
  typonymous homonym, *206*
  typotype, *206*
  typ(us), *see* type
    cons(ervandus), 163, *206*

ultrastructure, 41, 98, 105–6
uninomial, *206*
union, 90, 95
unitary name, 145, *206*
unite, *206*
units, measurement, 53
Universal Decimal Classification, 113
unpublished name, 125–6, 148, 155, 162
usage, 128, 130, 136, *206*

valid publication, 39, 41–2, 56, 91–2, 125–126, *134*, 138, 152–9
validly published name, *207*
validating author, 125, 160, 162
variant, *91*, 95, 99
  orthographic, 169, 173; *see also* orthography
variety, 38, 42, *43–4*, 45, 65, 133, 149–50
  chemo-, *44*
vernacular name, 173, *207*
voucher specimen, 25–6, 35–6, 76–7

xeroxing, 112

yoked key, *72*
younger synonym, 130, *207*